I've met many engineers who think they're writers. I've even met some writers who think they're engineers. I've never met anyone like Mike, though, who truly is both engineer and writer. He writes beautifully about his experiences, always keeping the spotlight on the people around him and their passion for helping make the world a better place every day.

Ben Walpole – ASCE News Senior Manager

Praise for *Bridging Barriers*

Thanks to God and to the Engineers Without Borders members, our water project and bridge are working well. We hope that EWB continues helping other communities that are in need, just as we were.

Don Rolando – Community Coordinator for La Garrucha

Mike's passion for serving others comes through in everything he does, and this book is no different. The challenges of these projects and the satisfaction of working with students come alive in reading this book. If you have ever thought of giving back or using your engineering skills to make the world a better place, read this book!!

Dr. Mark Federle, PE, CPC – Professor & Associate Dean for Academic Affairs, Marquette University.

Bridging Barriers provides critical information to help make global engineering work a success. It should be required reading for all students and mentors interested in understanding how personal stories, community needs, mentoring and engineering service all combine to make the world a better place. It inspires our next generation of ethical builders.

Dr. Daniel Zitomoer, PE – Chair and Professor of Civil, Construction and Environmental Engineering, Marquette University.

Mike Paddock's engrossing story is a firsthand, on-the-ground lesson in how communities can take charge of their futures against long odds and many obstacles. It's a must-read for engineering students who want to make their lives matter, and for service organizations and philanthropists who want their money to make a difference.

John DeDakis – Journalist, Novelist, Writing Coach, and Former Senior Copy Editor for CNN's The Situation Room with Wolf Blitzer

This book brings to life that engineering is all about people—not numbers and calculations. None of our classes provide the level of insight into the actual world of engineering as much as working on a real project that helps real people. Engineers Without Borders has given me my own looking glass into reality. This book is exactly what I was looking for to provide guidance and context for our EWB Project.

Alexandra Solecki – Marquette University Engineering Student

Rotarians have learned that drilling a well alone will not solve a community's water problem. It is essential to understand the dynamics of the community—how the well will change their lives and what it will take to settle into a new status quo. This book provides insights and lessons learned in community-building that will lead to the success of many projects around the globe.

Mary McCormick – Executive Director, Rotary Club of Milwaukee

In *Bridging Barriers*, Michael Paddock delivers three important stories: practical wisdom for engineers working on development projects in emerging areas; humanizing the situation of the poor in Central America; and paying tribute to a friend and mentor, the late Sir Michael Shawcross. It is lively, engrossing reading for people interested in any of those topics.

Avi Lank – retired columnist for the Milwaukee Journal Sentinel *and co-author of* The Man Who Painted the Universe

BRIDGING BARRIERS

How a Community Changed Its Future with Help From Engineers Without Borders USA Volunteers

---

Michael Paddock, PE, PS

Ten|16
PRESS

ten16press.com - Waukesha, WI

For information, please contact:

Ten|16
PRESS

www.ten16press.com
Waukesha, WI

Cover design by Kaeley Dunteman

AUTHOR'S NOTE:

This story is based upon actual events. For the sake of privacy, some names and locations have been changed.

DEDICATION:

I dedicate this book to Michael Shawcross, "Don Mike."
He was a man with a heart as big as Guatemala who mentored and showed me how to do development work around the globe.

"Thanks for Showing Us How To Help"

TABLE OF CONTENTS

Foreword
Bernard Amadei, Founder of Engineers Without Borders USA............*11*
Cathy Leslie, Executive Director of Engineers Without Borders USA...*13*

Chapter One:
Introduction..*15*

Chapter Two:
The Land and its People..*29*

Chapter Three:
The Troubled History of the Mayan People.........................*41*

Chapter Four:
The Importance of a Solicitation.....................................*59*

Chapter Five:
The Students' Bridge Design..*81*

Chapter Six:
The Construction Begins with a Surprise...........................*101*

Chapter Seven:
Hurricane Stan Disaster Response...................................*118*

Chapter Eight:
How to Pay for a Bridge..*132*

Chapter Nine:
The Bridge Construction..*143*

Chapter Ten:
The Landcruiser..*168*

Chapter Eleven:
The Request for Water – Water is Life..................................*174*

Chapter Twelve:
Designing and Funding the Water Project..................*184*

Chapter Thirteen:
Setting Up a Water Utility....................................*206*

Chapter Fourteen:
Building the Water System....................................*217*

Chapter Fifteen:
Water Utility Post Construction Coaching...................*233*

Chapter Sixteen:
Encouraging Community Leadership........................*249*

Chapter Seventeen:
Don Mike's Report Card after Ten Years...................*264*

Epilogue...*281*

Acknowledgments...*286*

About the Author...*289*

Foreword

As the founder of EWB-USA, it is my great pleasure to endorse the book of Michael Paddock titled *Bridging Barriers*. This book should come with a warning: reading this book will change your personal and professional life for the better, whether you are a professional engineer, an engineer in training, an engineering student, or not even an engineer. This book presents a clear and detailed case study of *how* to carry out community-based projects in the developing world context. It illustrates clearly that such projects are not just about coming up with technology to fix specific development issues. After all, if all development projects were only technical in nature, we would have solved them by now. Development work is foremost about people. It also involves dealing with multiple issues in an integrated manner under constrained conditions, whether they are cultural, geopolitical, environmental, social, etc. Finally, development work is context and scale specific.

Of course, this book includes engineering design and planning, implementation, monitoring and evaluation, and all the different components of project management familiar to engineers. This book is, however, more than that as it encompasses the human dimension of engineering at the grassroots level, community participation and empowerment, and more importantly, long-lasting friendship and compassion in action. This book is about engineering with a human face and involves bridging heart and mind, a challenge to most left brain-oriented engineers.

In this book, Mike Paddock, an imminent professional engineer and EWB-USA professional mentor, shares with us his story on how he got involved in several projects in remote communities in Guatemala. The book is about thorough engineering decision-making, collaboration with local communities, how to integrate lessons learned in various projects, and dealing with local cultures and different aspects of community dysfunctionality. In addition, this book is about a genuine friendship that the author developed over several years with an eclectic British expat, Don Mike. Lots of wisdom comes out of the story told by the author as he works side by side with Don Mike. This wisdom will be of great interest to students of engineering and those young people who contemplate becoming engineers. Wisdom is not learned in books; it can only be learned through experience in the field and in a hard way through several cycles of trial and error.

This book will also take you through the daily lives of remote communities in Guatemala, their dreams and hopes, their frustrations, and ultimately their hope to live with dignity and at peace. This book tells you a beautiful story that reads like fiction. Keep in mind, though, that it is a real story that one day may hopefully be converted into a motion picture.

I hope you will enjoy reading this book as I did. The development and humanitarian engineering world needs more books like the one you are holding in your hands and less uninspiring reports written by individuals who occupy the ivory towers of academia, NGOs, and international development organizations. Thank you, Mike, for showing the way and sharing your beautiful story with all of us.

Bernard Amadei, PE
Founder of Engineers Without Borders USA

Four months after I initially wrote the foreword and just before this book published, we find ourselves in a different world due to the COVID-19 pandemic. This pandemic reminds us that life is sacred and that access to health care and clean water shouldn't be taken for granted. As you read this book, think about the effort communities, such as La Garrucha, have made to increase access to health-care facilities and clean water and how that will better position them in slowing the spread of COVID-19. This pandemic is truly a test of preparation and of sustainable development solutions in communities as they brace to respond to an event like COVID-19.

After months of corresponding via email at night after our daily jobs were finished, I finally met Mike Paddock in, of all places, a hotel bar in Denver just before one of the early Engineers Without Borders USA conferences. Little did I know then that this was going to be a friendship that would take us around the world to many bars, debating points of engineering infrastructure, gender, culture, community impact, and just plain debating for the sake of debating.

Mike, to me, is an inspiration for how volunteers can do so much good in this world. Besides volunteering and working for EWB-USA, he has volunteered for Rotary, PAVA, Bridges to Prosperity, and the government of Guatemala. He has improved the lives of thousands, if not hundreds of thousands, of people who needed engineering infrastructure both before and after natural disasters.

But, Mike is not just any ordinary volunteer. He continually tries and tests to ensure that the infrastructure he provides is sustainable and doesn't create a dependency on Non-Governmental Organizations (NGOs) to come back and fix. He pushes the boundaries of every organization he works with in pursuit of continuous improvement and community impact. As one of our past mentors, Bud Ahearn, said, "You can't make an omelet without breaking some eggs," and Mike is a fabulous omelet maker!

This book is a story of one man's journey into volunteerism and how he has made a difference. As you read this book, you will be amazed at the incredible people that are a part of this real-life story. You will shed tears, you will find yourselves wondering how anything gets done, but lastly, you will be inspired to act and to make your own difference. Your challenge: find your passion and make your difference, don't just let the world pass you by.

Mike and I will find you at the next hotel bar somewhere in the world—doing good.

Cathy Leslie
Executive Director of Engineers Without Borders USA

CHAPTER ONE
Introduction

I slowly bumped along the roadway in the old, rusty pickup truck, navigating the route across the mountainous Guatemala Highlands road. It was November and my favorite time of year in the Highlands. The rainy season had recently ended, and the mountainsides were lush and green with their fresh new growth. Bright mountain flowers greeted me along the route with a new surprise around every turn. The crisp air smelled fresh and clean as if it was washed by the recent rains.

But, there was a large knot in my stomach. I was worried and nervous. It made me think back to another time when I felt this way. The time when I sat nervously for my Professional Engineering Board Exams, some thirty years earlier. That was a major career milestone that tested my life's work. And this trip to Guatemala now felt like another major exam. I was traveling to visit the community of La Garrucha ten years after the Marquette University Chapter of Engineers Without Borders USA (EWB-USA) completed its program there with the assistance of Rotary International.

The program consisted of building a bridge and potable water system to improve the health of the community. I had been the program's Lead Engineer and tasked with mentoring the student teams who had worked so hard on fundraising, designing and building the projects with the community. Now, ten years later, I was part of a review team that would spend the better part of two weeks documenting the lessons learned from the program.

What would we find?

I always had been a believer that infrastructure development work can have a lasting impact if done correctly, but there was also a small voice in the back of my head expressing doubt. Would I find a successful, functional system? Or would I find another failed infrastructure project like so many that I had been asked to fix over the decades?

The potential consequences of failure were serious. The old saying "Water is Life" is true, but what is not said is that contaminated water can lead to death. Drinking water systems that are not properly designed and maintained can become breeding grounds for contamination and disease that is spread across the community. Failed infrastructure projects also disrupt the social fabric of the community. Leaders who advocated and supported the project can be ostracized and the community can become risk adverse, not willing to attempt another improvement for fear of failure. So the stakes were high, indeed.

Surely the wisdom of my mentor and friend Michael Shawcross, or Don Mike as he was called locally, had guided the community to success; but my doubt remained.

The idea of doing the La Garrucha review had come from EWB-USA's Executive Director, Cathy Leslie. Cathy is an accomplished engineer and the heart and soul of the EWB-USA organization. Like all engineers, she is committed and lives by the engineer's oath of "Protecting the health, safety and welfare of the public."

Her commitment to service engineering started upon her graduation from Michigan Tech University when she joined the Peace Corps where she designed and built water systems in Nepal and then never stopped. Cathy had been one of the founding members of Engineers Without Borders USA in 2003 and had faithfully guided it along its way as the executive director since 2004. She is a strong leader with vision who people cannot help but want to follow.

Cathy and I had celebrated another year of EWB-USA's existence at the annual conference with a toast to the organization a few months earlier. We had grown to become good friends over the years, enjoying each other's company and sharing our life challenges and accomplishments.

As our glasses of beer clinked together while we relaxed at the hotel bar, she asked, "So, what percentage of EWB-USA's projects do you think would be considered successful?"

We both knew the statistics that nearly half of all water projects around the globe had not been able to sustain themselves and failed.

"Do you think EWB-USA's projects are all that different?" The look in her eyes told me that she truly wanted to know. Her thirty years of engineering experience had taught her to ask the right questions and now she looked back at me over the top of her glasses seeking answers.

As do many engineers, she possesses the desire to identify and solve problems that help people and she was determined to find some answers to this question. It is the same quest for improvement that had pushed her to move the organization from its fledgling status to a powerful nonprofit with over 15,000 members in 300 chapters that complete hundreds of projects each year.

"I'd like to think that our projects are doing better than most, but I have to admit that I don't know that for sure." I replied.

She continued to press. "I really want to know what the impact of our projects are and how our communities are doing. I want to know the _real_ story - beyond the number of beneficiaries, monitoring and evaluation data collected one year after a project is completed."

She looked me in the eye as she continued, "As engineers, we know that infrastructure projects need years to be able to demonstrate sustainability. Go to one of your early projects that are now ten years old and answer the question. I can't imagine that we won't be a better organization because of it."

I knew she was right as I sipped my beer and peeled the label. I had always wanted to follow up on the La Garrucha program and had always found excuses to not do so. It never seemed to be the right time, but the reality was I was afraid of what I might find. Now I had run out of excuses as Cathy stared back at me over the table. Despite my fidgeting, it was clear that she would not allow me to change the subject.

"Ok, I'll do it." I said quietly before taking another drink of beer for courage.

She smiled.

As I continued to drive in the Highlands, a loud horn blared, startling me out of my reverie. A large bus raced past my pickup. Buses are the transportation lifeline for the Highlands rural communities for education, health care, and markets. But they also help people with other important duties such as grandmothers visiting their grandchildren on a Sunday afternoon or a young man nervously traveling to visit a lovely young lady in the next town for the first time.

Foreigners refer to them as "chicken buses" reflecting the varied cargo that they move. Not only is it possible for a chicken to peck a hole in your shoe while riding on the bus, sometimes an entire adult pig has been seen on top of the bus – "hog tied" to keep the frightened traveler from jumping off. I remembered the time when I rode next to a lovely woman who was traveling to the market with an entire hive of bees to sell. From then on, I chose my seat more carefully.

The bus is a good sign. Ten years ago, no bus service existed in this area. Surely the vehicle bridge that was built across the Rio Motagua as part of the program must be in use. I was also impressed by the condition of the roadway, which was vastly improved since I had last traversed the route years ago. Even though it was the end of the rainy season, the pickup was able to easily navigate the route without using its 4-wheel drive. My hopes rose, but I was still concerned. What would be the condition of the water system?

Drinking water systems require regular maintenance and upkeep that, in turn, require a utility with fees. As with any utility, my mentor Don Mike had always said the questions are simple:

How do you generate the bill?

How do you collect the bill?

What do you do when someone does not pay their bill – especially if that person is your grandmother?

Three simple questions that water utilities have struggled to find the answers to all over the planet. Time and again I had been asked to repair water systems built by other organizations that had been unable to keep up the repairs. I had come to refer to the process as the "Water Death Spiral" as users refused to pay for unsatisfactory service, which in turn, resulted in less money available for the needed repairs. The cycle would continue, resulting in even poorer services that more people refused to pay for, finally ending with the system spiraling to total failure. Would the La Garrucha water system suffer a similar fate?

I knew these were the difficult questions that many ask about international development work in general and about work that utilizes volunteers in particular. On a personal level, I was also hoping for some validation of the over 20,000 hours of volunteer work I had invested into engineering service projects around the globe. Was the effort really worth it, or was I simply doing "push-ups in the corner" and wasting time?

Now I safely crossed the concrete bridge that spanned the raging river twenty-five feet below. I continued on the roadway up the other side of the valley and reached the school building where I and the rest of the review team were greeted happily by my old friends Rolando and his mother Gavina. Both looked exactly as I had remembered them nearly five years earlier.

Rolando wore his cream-colored cowboy hat that was perfectly creased. He had his hand-woven woolen bag slung across his shoulder, hanging gently along his side. A smile filled his face which is framed by a well-kept beard. His eyes danced with excitement. I swear that he hadn't aged a day as he shook my hand vigorously.

Gavina wore a neatly pressed apron with a colorful blouse underneath. Her apron never seemed to be dirty, despite the fact that she always wore it to keep her blouse clean. Her hair had a few more gray streaks and it was pulled back into a tight bun as I had remembered. She tipped her head slightly to the side and flung open both her arms, inviting me for one of her patented bear hugs. I could feel her energy flow into me as she squeezed me and gently rocked from side to side.

We all agreed that it had been way too long since we had seen each other.

"Could it really be five years?" Rolando asked. We also sadly remembered that it had been four years since the passing of Don Mike. Everyone shook their heads in disbelief as the friendship seemed as fresh as the day we had last seen each other.

Gavina then placed her hands on her hips, scolding me since we had only given her two days' notice for the visit. The surprise visit was intentional on the review team's part to keep the community from racing to do any repairs that might skew the review results. The community had clearly rushed to empty a room in a home where we could sleep while doing our work. The concrete floor was spotless, and the walls were covered with past calendars that had been saved for their landscape pictures. A few plastic chairs were placed around a handmade wooden table in the middle of the room. To greet us, it had a small glass placed on it with a few wildflowers carefully arranged. I could also not help but smell the steaming corn tortillas and the freshly brewed Guatemalan coffee that had been prepared for us.

The community had self-identified a health crisis which was the reason for them seeking help over a decade ago. Two to three babies and one of the community's young mothers were dying each year due to fatal deliveries. Gavina and the other midwives had reached their breaking point and tirelessly advocated for change. They knew what to do and how to help their patients. But how can one do an adequate job as a health care worker without having access to a hospital and clean water? That was the problem that the EWB-USA / Rotary program was tasked with helping the community solve.

The next morning, the review team began its work accompanied by members of the community. The review team visited each of the homes of the 1,500-person town called an *aldea*. We trudged up and down mountain paths to interview the families about their health while performing water quality tests and measuring the flow rates at each water tap. The school and health directors were interviewed, and the records compared to the baseline data collected before the project was started.

The results were amazing.

The bridge and water systems were the catalyst of change that sent the community on a path of continuous improvement. They had eliminated maternal deaths, reduced the infant mortality rate to one in ten years and reduced monthly school absences from over 300 to just two.

As I prepared to say goodbye, I wandered over to the local soccer field. There, twenty-five youngsters were playing a game under the watchful eye of ten mothers. I thought of Don Mike and smiled, thinking, *Thank you. If you had not come, these people would not be here.*

I recognized most of the mothers who were at the field cheering on their sons and daughters as they waved to me and I waved back. They were not just statistics – they were real people, people who now had hopes and dreams of a better, healthy life.

As I left and the old pickup groaned on its way up the mountain, I realized that the knot in my stomach was gone. I had found my answer. But then I realized, the important question was why and how these changes occurred?

I knew it was largely due to the wisdom and guidance of Don Mike. Don Mike had happily shared his wisdom and lessons with me over two decades. Lessons that had also largely helped shape EWB-USA into the organization it is today. Before meeting Don Mike, I thought I knew the answers to community development. Now I realized that I didn't even know the questions. I knew that these lessons must be passed on to others who wished to help with development work. But how?

I then remembered the words of Don Mike. "When one is perplexed with who is the right person to answer a challenge, it is sometimes best to start with the person in the mirror."

I swallowed hard and decided to write down the story of EWB-

USA's La Garrucha program to help others learn the lessons from Don Mike.

...

Michael Shawcross was a British citizen who had come to Guatemala at the age of twenty-nine in the early 1970s and never left. Born on September 23, 1941, he was a stout man of less than average height. He always wore a British flat cap and sported a long, white, unkept beard. His thick British accent captivated his audiences and he was never hesitant to utter a few swear words now and then to get his points across. He loved a drink, especially scotch, but was more than happy to sample the local moonshine that the Mayan people make, called *cusha*.

After completing his studies as a young man, Don Mike's wanderlust pushed him to take a hitchhiking adventure from England to China with the equivalent of only twenty dollars in his pocket. He made it only to India before illness sent him packing home. When he arrived unannounced on his mother's doorstep on Christmas Eve, she gave him a huge hug and the tears flowed. Then she demanded that he wait on the doorstep while she gathered a change of clothes for him. Those dirty clothes would not be allowed to enter her house and would go directly into the garbage. Only her son would be allowed to enter if he promised to pass directly to the shower. After he had scrubbed away the dirt and smell accumulated over the months, one of the best Christmas Eve celebrations the family ever had ensued.

Following his adventure to India, Don Mike's wanderlust took him to Canada where he worked in the old Northwest Territories mapping national parks and caves. He loved caving and everything caves contained. He loved the adventure, the challenges, the fear and the joy of exploring. He would later say that he was amazed that the Canadian Government actually paid him to explore caves – he loved it so much he would have done it for free.

It was his love of caves that brought him to Guatemala, but it was his love affair with the Mayan Highland People that kept him there. He so admired their history, culture and commitment to community that he could not help but love them. He found it easy to help these people who also worked so hard to help themselves.

He wrote one of the first travel books of the country and opened a bookstore in the Guatemalan city of Antigua. The bookstore was a perfect fit for him, and he would always refer to himself as a "book man." The store not only contained books, but maps of the area and his beloved caves. Soon cavers from around the world would visit his store to gather information on caves and areas to explore. He would collect rare literary works in Guatemala and sell them to The Library of Congress and many universities around the world to help pay the bills. This allowed him to explore the far corners of Guatemala, much of which he did by foot.

On February 4th, 1976 a massive 7.5 magnitude earthquake shook the country as the North American and Caribbean Continental Plates slid against each other. The Mayan adobe homes were no match for the earthquake's power. Nearly everything was destroyed. More than 100,000 people were killed or injured and nearly 1.2 million[1] found themselves homeless.

Don Mike was thirty-five years old at that time. He and his friends saw the need and gathered what funds they could to help the Mayan

---

[1] Olcese, Orlando; Ramón Moreno & Francisco Ibarra (July 1977). "The Guatemala Earthquake Disaster of 1976: A Review of its Effects and of the contribution of the United Nations family"

farmers. They knew that the farmers would need new seed stock and tools to replant their fields before it was too late in the growing season. If the year's crop could not be salvaged, the earthquake disaster would be followed by an even greater disaster - famine.

Don Mike learned that working with the Mayans fit him perfectly. He and his friends saw the need in the communities and knew education was the key to an improved life. Hence, PAVA was formed as an official Guatemalan non-profit organization whose work would focus on education in the rural areas and continues today.

During the 36-year Guatemalan Civil War from 1960-96, or "The Violence" as it is referred to by the Highlands People, Don Mike saw the need caused by the war that destroyed infrastructure and interrupted the health care and educational services in the most affected regions. As was his nature, Don Mike ran towards the war, not away from it. He formed a non-government organization (NGO) at the age of forty in the Ixil Triangle region and began building and operating schools and water systems.

It was during this time that he sharpened and perfected his development skills. Fortunately, he was always more than willing to share his lessons learned with others who wished to work in development. For over a decade, Don Mike served as the local coordinator for EWB-USA. He voluntarily coordinated projects and EWB teams, always refusing compensation for his efforts.

It was through these years of selfless giving that he earned the respect of the Highlands people he so loved. His efforts were also noticed by the British Ambassador, resulting in Don Mike being awarded the Order of the British Empire (OBE), which he graciously accepted from Prince Charles. True to his heart, he lobbied Prince Charles as the medal was pinned to his chest, saying, "Thank you,

Your Highness. Now about the British support for the education of the people of Guatemala . . ."

Upon his death at age 73, his sister unexpectedly received a solicitation while she frantically looked for a cemetery that would bury a British Citizen in Guatemala. The solicitation contained countless thumb prints, which act as signatures, from the communities in which he had worked. The solicitation asked if his sister would allow them the honor of burying Don Mike in the Highlands cemetery in Joyabaj. Of course, she agreed and to this day if one visits the grave, you will find it well kept and adorned with wildflowers from his beloved Highlands.

I am a professional engineer and surveyor who met Don Mike twenty years ago. At age thirty-one, I survived a near-death experience with cancer after being given less than six months to live. Finding myself quite unexpectedly in full remission, my wife, Cathy, and I

Don Mike's grave located at the Joyabaj Cemetery.

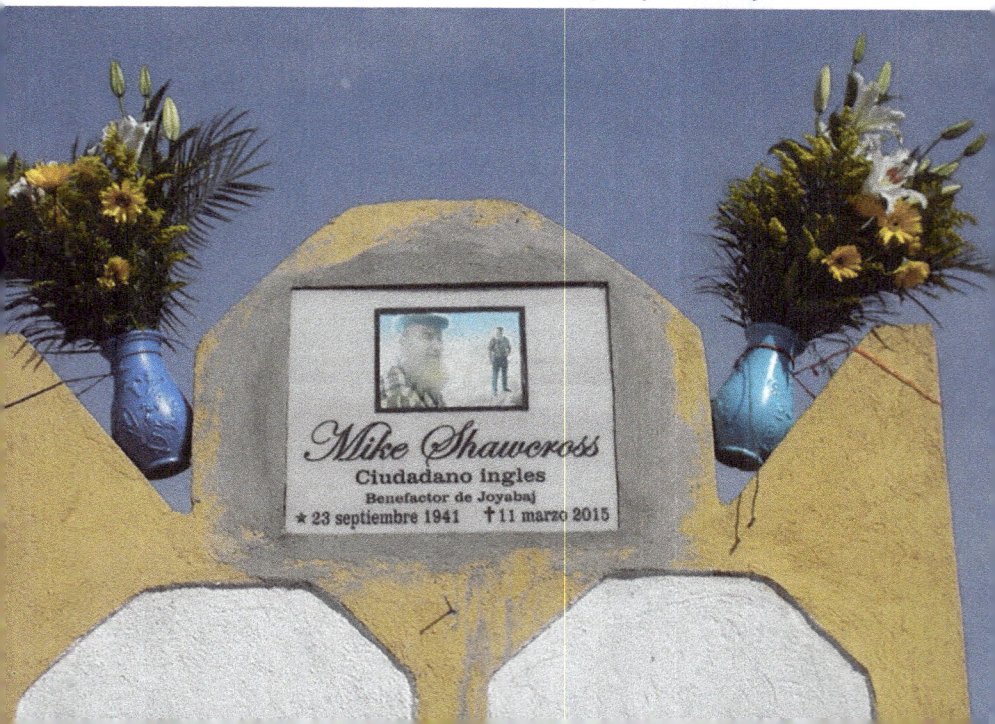

plotted out our future to let me use the gift of life to help others around the world. That is how this civil engineer who managed multi-billion dollar freeway projects in Wisconsin became a full-time volunteer engineer for EWB-USA, traveling to some of the most impoverished and devastated places on five continents.

While on one of these assignments in Bolivia, I met Paul Johnson, who happened to also be an engineer from Wisconsin. He was waiting patiently for the survey equipment that I had borrowed from a missionary for his own project. I was a bit nervous meeting him at the local ice cream store, as I was nearly a week late in the delivery of the equipment. But he welcomed me with open arms and we soon were discussing engineering projects – a habit that most of us engineers possess. Through these discussions, he told me about his work in Guatemala with a crazy old Brit named Don Mike and I was hooked.

As Paul tells the story. "You may have gotten the survey equipment for an extra week in Bolivia, but I got you to work in Guatemala for the next twenty years. I think I got the better end of that deal."

When I had met Don Mike on my first Guatemala project, he was depressed. The love of his life, a wonderful Mayan woman, had moved away and become engaged to another man. Even though they had worked together for years, Don Mike had never gotten up the nerve to tell her how he felt. Now she was gone. He never fell in love again.

Don Mike threw himself into his work to help the Mayan people as a distraction from his sorrow. I always felt that I was also a beneficiary of his broken heart as he took me under his wing. He was happy to mentor me and share his wisdom from his decades of experience.

When I entered Don Mike's home for the first time, I was amazed – it was literally packed with books and publications. The floor was covered with stacks and stacks of material to the point that only a small path was allowed for anyone to walk through the room or corridor. This was no ordinary bachelor's pad; it was truly a mess. One particularly large mound caught my eye, and then I realized that it was a dining table that was completely engulfed in books. A small area of the tabletop, no larger than a plate, was reserved for Don Mike's dining. The top of Don Mike's refrigerator had over a dozen gallon plastic gas cans on it. They were filled with *cusha* that had been gifted to Don Mike. Each had written on it the names of the communities that had provided the gift.

"How about a little eye opener to get the day going?" Don Mike had asked with a grin as he searched for a clean glass.

I then noticed a wooden plaque nearly completely buried in the debris. It was from the United States of America Ambassador Thomas Strock. After some prodding, Don Mike reluctantly shared the story of how he was invited to a gala in the capital – the type of event that he hated. As he stood on the steps of the hall, he changed his mind about attending. He knew that he probably would not fit in and he wouldn't be much company to those around him, so he returned to his home. The next morning, he was informed by his friends that the event was held in his honor for his development work during the Guatemalan Civil War. They knew he would not like the attention, so they had planned to surprise him with a wonderful evening. "I guess that is one event I should not have skipped." He said as he looked down at his glass of *cusha*.

The words inscribed on the plaque are:

"Thanks For Showing Us How To Help."

Don Mike showing off the radishes from a family garden near Tecpan.
Photo Credit Glenn Kozicki

CHAPTER TWO
The Land and its People

"It all starts with the land.
One must fully understand the land
and its relationship with the people
that inhabit it – what it provides freely
and what challenges are posed to those
who must live in concert with it."
- Michael Shawcross

One of the first lessons Don Mike taught to me was that the environment of any community shapes its behavior and culture. The community has lived with the risks of its environment for generations and it has become ingrained into its society. Anyone working with the community must understand these risks, the stresses they inflict on the people and the history of any conflict they may have caused.

The Land

The community of La Garrucha is nestled in the heart of the Guatemalan Highlands, just south of the mighty Rio Motagua, Guatemala's longest river. The Country of Guatemala is uniquely located at the point where three continental plates converge – the Pacific Plate along its western coast; and the North American and Caribbean Plates that meet along an east-west fault that splits the country in two like a cleaver.

The continental plates float upon the surface of the earth and are always moving. Like massive titans, they jockey for position against each other, sometimes gently and sometimes in massive displays of violence and force – earthquakes and volcanos. For centuries the North American and Caribbean Plates have continued this struggle against each other in a constant battle for space with the Caribbean Plate muscling itself to the north against its neighbor. Over time, the Caribbean Plate has been pushed under the North American Plate and this action has created the massive mountain ranges on each side of the fault line that rises over 3,000 feet into the heavens.

The fault also brings a danger of earthquakes. The most significant in the people's memory was the event of February 4th, 1976. The 7.5 earthquake shook the land starting at three in the morning, awaking Don Mike. The force of the earthquake overwhelmed the simple adobe home construction and virtually leveled every building in the Rio Motogau valley. It is estimated that 100,000 people lost their lives or were injured, and 1.2 million people were left homeless.[2]

Don Mike explained that although these events occurred nearly fifty years earlier, the memories remain so painful and vivid that the people live their lives with a wary eye towards the fault and its dangers. "You don't live through a disaster like that without the memory being burned into you for the rest of your life," he said stroking his beard remembering those dark days.

Any infrastructure project must account for the earthquake risk that is inherent to the region. Don Mike explained how the communities mitigated the risk of an earthquake with different levels of protection. Homes needed to be built to not collapse on those who may be inhabiting them. Schools needed additional protection as they will become a place of refuge for the community in the event of a disaster. Bridges also needed to be built with a high degree of protection as they will be the lifeline, providing vital access for relief supplies and assistance in the event of a disaster.

The geology is also the source of the incredible beauty of the Rio Motagua valley and Don Mike reveled in its glory. The active fault has made the new mountains with jagged and sharp rock outcrops that jut up into the sky creating a rugged beauty. The slopes of the mountain are steep due to the constant uplift of the plates, and boulders are constantly tumbling down the mountains as they lose their foothold.

Along the fault line runs the mighty Motagua River. It flows along a steep and winding path at the base of the two mountain ranges

---

[2] Olcese, Orlando; Ramón Moreno & Francisco Ibarra (July 1977). "The Guatemala Earthquake Disaster of 1976: A Review of its Effects and of the contribution of the United Nations family"

providing its blue-ribbon accent to the landscape with whitewater streaks as the river moves through its rapids. Some tall pine trees gain a toehold on the steep slopes and add a green accent to the mountainous landscape.

This pattern is only interrupted in areas where a slightly reduced slope allows a farmer to walk and carve out a small field of corn and beans using his *hacedone* or hoe – just like his father, grandfather and great grandfather did. The corn rows are contoured horizontally across the slope which provides a ringed texture to the landscape that blends into the natural beauty as if it were intended by nature.

The beans and corn are planted together in the same field in a practice called polyculture. They work together with the beans providing important ground cover to protect the soil's moisture while the corn provides needed shade for the beans. As the fields grow, first the beans appear with their dark green accent and then the corn begins to tower above with its lighter green color. This provides a constantly changing color pallet to the view that marks the passing of the seasons.

As I viewed the valley for the first time, it took my breath away as the image is striking. Add to the view a volcano or two in the distance and the view becomes mesmerizing.

The other senses are not to be denied. Fresh air fills the lungs, body and soul as it is breathed in. Just before planting, the faint smell of smoke is in the air as the fields are burned to release their nutrients back into the soil to feed the next season's crop. When the fields are hoed and readied for planting, the smell of fresh earth adds its accent to the crisp mountain air.

The soil is volcanic in nature and rich in nutrients that feed the vegetation that grows upon it. Its texture is soft and warm to the touch accented by the volcanic ash that slides easily between one's fingers.

Many Mayans still farm without shoes and this allows the soil to squeeze up between their toes, literally connecting the farmer to the earth that provides him and his family life.

The sound of the river is a constant backdrop and lulls one to sleep each night. It is interrupted each morning by the crowing of the roosters as they rudely announce the new day. As the sun warms the valley, a breeze is pushed up the mountains through the pine trees creating its own accent to the ensemble that relaxes the listener while viewing the landscape. Don Mike never tired of the incredible beauty the Highlands offered and would insist on stopping from time to time to simply bask in its glory. "Sometimes you need to stop and pinch yourself to make sure you are not in a dream," he said.

The Motagua River Valley at La Garrucha

The Water

Water is necessary for all life and its cycle needs to be incorporated into any project. For six months a year, rains fall with regularity and water can be harvested for drinking and agricultural purposes with relative ease. Although an outsider may get frustrated by the daily rains, the Mayans welcome it with a smile and the old saying that "water is life." During this time, one can almost see the corn growing in the fields as it drinks from the rain each evening and basks in the morning sun.

But during the six-month dry season, barely a drop of rain falls, causing a shortage of water to the region. The corn and bean fields have adjusted to this challenge. The seeds are planted at the end of the dry season, allowing them to sprout in the first early rains. The crops grow throughout the rainy season and are harvested as the rains end. This allows the fields to simply lie dormant over the long months without rain.

But the people cannot lay dormant and must have water year-round to sustain their lives. As springs, rivers and lakes dry up the people are forced to travel longer and longer distances to find life-sustaining water sources, carrying the water to their homes on their backs. This annual struggle is part of the Mayans' lives and has been ongoing for generations.

The inherent challenge of drinking water is enhanced by the continental fault that pushes one plate under the other, causing ground water to be nearly impossible to access north of the fault line. In this zone, no springs exist, and a well is dependent on luckily finding a fracture in the volcanic rock more than 600 feet below the surface that might provide some life sustaining water.

Don Mike explained to me that even if one is lucky enough to drill a well that finds water, a costly generator and diesel fuel are required to pump the water when grid power is not available. The cost of the fuel to power the generator can easily consume one third of the community's entire income, making a well an unsustainable solution in many cases. Unfortunately, this has been proven time and again in the region as wells with diesel generators sit idle due to a lack of sustainable funding from the communities for fuel.

Those living south of the fault are more fortunate as the steep terrain allows the groundwater to flow out of the ground at numerous locations in the form of natural springs. These spring sites have been well known to the people for generations and are passionately protected due to their importance. Water is not only needed to nourish one's body but also to sustain any livestock such as chickens and pigs that a family may have. As the old saying goes, "whiskey is for drinking and water is for fighting." The access and use of springs have been a source of conflict for centuries and continues to be a flash point between those who have access to clean water and those who lack it.

The Rio Mogatua

The beauty of the landscape also brings other dangers. Not only the danger of rock slides, but the steep mountainous sides cause any rain to rapidly run off to the river below. The steep grade of the mighty Motagua River swiftly carries its waters towards the Atlantic Ocean at a high rate of speed and with incredible force. This is the source of flash flooding that can cause the Motagua River to quickly rise and transform from a lazy, wandering river to an ugly torrent that will sweep away anything in its path in a matter of a few seconds. This is especially true during the hurricane season that occurs in the fall of each year when the river's force can easily move boulders the size of trucks.

Over the centuries, many a would-be traveler would attempt to cross the river by swimming its temptingly narrow width, only to be swept away by its strong and unrelenting current. The Mayans have come to give the mighty Motagua the nickname "The Assassin" for its reputation of being a killer who strikes without prejudice or discrimination. Despite the many stories of strong men being swept away and leaving behind widows and children, men continue to test their skills at attempting the crossing despite their better judgment.

At one location along the river, the land has provided two natural rock outcrops that squeeze it together, making a rock canyon. This natural phenomenon has been known to the Mayan traders and has been called "The Crossing" or "*La Garrucha*" in Spanish. For centuries, traders have been funneled to the crossing as they moved their goods of salt, obsidian and other precious goods across the Rio Motagua. As the Quiche Mayan People located to the north traded their precious goods with the Kaqchikel Mayan People to the south, it was only fitting that a community was formed to the south of the crossing called La Garrucha.

Don Mike explained that the Mayan people differed from their Mesoamerican neighbors to the north (Aztecs) and south (Incans) in that they are not one united nation. They are made up of twenty-one separate nations that fearlessly maintain their independence with each having its own language and customs. The Rio Motagua has served as the natural border between the largest two Mayan Nations, the Quiche to the north and the Keqchikel to the south.

Of course, trade between the nations has always been an important part of society, but disagreements inevitably occured, and grudges are slow to die, so peace between nations was always a fragile state when it did exist. Although the Rio Motagua was always a barrier to free trade, it also served as an important boundary of separation during times of conflict between these Mayan powerhouses.

The Mayan People and Culture

The Mayan are generally small in stature, with lean muscular bodies that have been honed from the physical labor demanded of them. They are extremely friendly by nature but reserved in their expressions and generally speak quietly. Loud talk and flamboyant gestures are rarely seen and considered rude.

As reserved as they are with their speech, their dress is something completely different. The clothing is woven with bright colors and patterns that make any community gathering a beautiful ensemble of color. Each community has its own distinctive pattern and style to its weaving so anyone can easily recognize the person's home town simply by their clothing. On market day when large numbers of people gather, the colors and patterns dance for one's eyes like a kaleidoscope.

Within each Mayan nation, a strong sense of community exists. Since the early times, each family is expected to work on public works projects and programs that benefit the entire community. Although the work is not paid and voluntary, participation is not optional. All are brought up in the culture of participation in projects to benefit the entire community. The Mayan culture has adapted to provide resilience and sustainability in this land of challenges by relying on the community's overall strength and diversity. When one family is down on its luck, another is there to step in and pick up the load to help sustain the community's fabric.

The roots of the culture are based in subsistence agriculture – primarily the fields of corn and beans. It is believed that corn, or maize, was first cultivated by the Mayans' ancestors 7,000 years ago and has since spread around the globe as a food staple. The corn is typically soaked in lime water, boiled and ground to a paste called *masa* that is patted into neat, round tortillas. The making of the tortillas is an important activity for the community's women. Young girls, mothers and grandmothers stand around the fire together patting out the tortillas while they share the news of the day. The pat, pat, pat of their hands somehow always seems to be in rhythm and somehow never disturbs the conversation and laughter of these craftswomen.

A young tortilla maker processing the corn paste called masa.

The tortillas are cooked on the fire under the watchful eye of the ladies who carefully turn them with their bare hands. Their hands have grown thick calluses which over time protect them from the hot tortillas that easily burn any newcomer who tries their luck at the craft. It is not uncommon for a farmer to eat as many as forty corn tortillas in any given day and every meal would be incomplete without tortillas at the table.

Don Mike loved corn tortillas, especially those that had been cooked a bit longer on the fire resulting in a crunchy tostada. He would gently rub some salt on the surface and fawn over them, much to the delight of the cooks. Piles and piles of tostadas would be delivered to him during a meal and he was careful to always compliment the cooks, who would giggle and smile as they patted out another stack.

Don Mike explained that, in fact, many Mayans believe they are made of corn. Ancient legend has it that when God made man, he started with stone. Soon he found that the man of stone was unable to move, so he switched to soil. But the soil was brittle and was easily washed away by the rain so another material was needed to form humans. This is when God made corn and men and women were able to be formed into the beings we know today. When a Mayan says they are "People of Corn," they literally believe that their bodies are made of the grain.

Don Mike described that, as with all subsistence farming cultures, aversion to risk is a way of life. Sticking with the "tried and true" methods of one's grandfathers is the norm and deviations must be very carefully considered. New methods need to be tried in moderation and managed throughout a community because bold, new ideas that are tried and fail may result in harm or the destruction of the entire community.

These decisions to try something new must be discussed in an open format, together as a community. Any risk by an individual is a risk to the entire community's sustainability. Decisions need to be carefully weighed and considered with time given for plenty of contemplation, discussion and comment. Mayan people are generally reserved, and the discussion is usually quiet in nature with long pauses between expressions. Patience is key as it is important to hear everyone's voice and consider their point of view, as it is through this process that communities have avoided devastating failures that would cause the entire community to suffer.

Don Mike fully understood this aversion to risk and the need to take the time for everyone in the community to discuss and accept any proposed change. Unlike a western culture where the majority rules, the Mayan way is that even a few dissenters can block any proposed plan. Care must be given to everyone concerned to gain their acceptance, if not the approval of the plan. Many an outsider has been frustrated by the time and number of meetings needed to gain community acceptance, but Don Mike had grown to embrace the process.

"You cannot do any project until everyone in the community is ready, and not a minute before," he would say. "The community's decision-making process does not change for funding cycles, elections or other outside influences. It moves at the speed it needs to depending upon the community."

Don Mike with the author at the Chixquaina Bridge inauguration.
Photo Credit Mincho Ortega

CHAPTER THREE
The Troubled History of the Mayan People

"One must listen to the stories and tragedies of the past, even though they are painful. We will never be able to fully comprehend the impact it has had on the people, but it is important to try to understand the community through its eyes and feelings."
Michael Shawcross

It is important for anyone working with the Mayan people to understand their struggles and Don Mike was nothing short of a historian of the Mayan way of life. During long truck rides together, Don Mike would share his knowledge with me, explaining the facts as he knew them and his own view on the lasting impact it had on the communities. He knew that the historical context of the culture was important when working with any Mayan community and he was happy to educate anyone who was willing to learn.

He would say, "If you don't attempt to understand the people's journey, struggles and pain, you will never know the way to work with them."

One of places he revered most was a mural outside the community of San Martin Jilotepeque. The mural was painted by members of the Mayan community and beautifully depicts their history and struggles over the centuries. A single-phase on the mural sums it up well, and Don Mike teared up every time he visited it as he emotionally read, *"They burned our trunk: they cut our branches: they took our fruits: but they could not take our roots!"*

Mural outside the community of San Martin Jilotepeque

The trouble began for the Mayan in 1524 when Spanish Conquistador Pedro de Alvarado arrived in the region and quickly formed an alliance with the Kaqchikel Maya nation – an alliance the Quiche Mayans have never forgotten, even 500 years later as they viewed the Kaqchikel as traitors. Alvarado and the Kaqchikel Mayans systematically defeated each of the other Mayan nations including their main rivals, the Quiche. Unfortunately for the Kaqchikel, Alvarado then turned on his ally and defeated them allowing himself to become the supreme ruler of the region.

Defeating the Mayan was not enough for the Spanish as they systematically destroyed the Maya culture burning ancient Mayan texts and destroying their temples. The land was divided up among the victors and their friends, many who would never see the land that they owned. The people were forced to work on these Spanish plantations, called *fincas*. In the mountainous regions that were not suitable for largescale farming operations, large *fincas* were given to Spanish dignitaries and they were allowed to tax the Mayan inhabitants in exchange for converting them to Christianity.

This *"Mezos"* system continued for hundreds of years where the Mayan people worked in near slave-like conditions for the elite land holders. Mayans were not allowed to own land and were widely discriminated against and even considered subhuman by some. This is stunning when one studies the amazing ancient Maya culture and its technology advancements including the Maya calendar and astronomy.

Unfortunately, these practices eventually led to a grueling thirty-four year war referred to by the Highland people as "The Violence" that lasted from 1962 to 1996. A Truth Commission was formed out of the June 1994 Oslo Accord as part of the peace negotiations led by the United Nations. Don Mike explained that it is formally defined as a "historical clarification commission" indicating a focus on investigating the underlying causes and evolution of the conflict.

The Truth Commission's documentation of specific cases started with the events of January 1962, when the armed insurgency began, and continued through December 1996, when the conflict came to a formal end. The commission's analysis extended to a consideration of the historic and structural roots of "The Violence" changes and is widely regarded as the best source of information on The Violence. It should be noted that some believe the commission's report is one sided and does not fully represent the perspective of the army. The Commission began its work in 1997 and submitted its report in early 1999 as it was limited to only two years to do its work. In documenting and analyzing past abuses, the commission tried to explain how Guatemala came to experience the longest armed insurgency in Central America and the most violent conflict in recent Latin American history.

Don Mike never viewed the Truth Commission to be a complete or a fully accurate account of the events but considered it to be the best single source available. "How could they uncover decades of mistreatment and abuse in only two years?" he asked. He experienced The Violence firsthand, having been shot at by both the Army and guerrilla groups on various occasions.

At a few sites, he was the first person to visit a community after a massacre had occurred. These were memories that even Don Mike found difficult to share. For years after the peace accords, Don Mike worked tirelessly to find those responsible for several of his friends' deaths, spending countless days in the field investigating leads to the crimes. He was even able to eventually bring some of the murderers to trial and conviction after uncovering incriminating evidence. His accounts of The Violence were always filled with emotion as he vividly remembered with sadness and anger that terrible period of Guatemala's history.

Don Mike read and studied the twelve volumes that made up the "The Guatemalan Truth Commission Report" along with *Memory of Silence, The Guatemalan Truth Commission Report* edited by Daniel Rothenberg which, along with his personal experiences, formed the basis of his account. The documents provided important information on many of the communities where he worked and helped him understand the context they were set in. In many cases, the people found the memories too painful to talk about and simply wanted to put the past behind them. It was only after the reports were referenced in personal conversation, that the stories were shared with outsiders.

Don Mike explained that, as the Truth Commission also found, understanding "The Violence" requires a deeper look at a number of factors within Guatemalan society and history, particularly the nation's severe poverty and profound structural inequality. Going all the way back to the conquistadors, a small elite group identifying itself as ethnically and culturally distinct has long controlled most of the country's land and wealth. From the colonial era on, Guatemala's social and political system has supported this gross inequality, linking the economic dominance of a minority with systematic discrimination against the majority, especially the nation's Mayan population. The country long relied on a series of repressive laws and regulation that, when challenged, were unfortunately backed up by state violence.

Even now, more than two decades after the negotiated peace, the nation is among the poorest and most unequal in the hemisphere. Over half of all Guatemalans live below the poverty line, with fifteen percent living in extreme poverty. More than forty percent of children under five are chronically malnourished, and the country has some of the region's worst social statistics regarding health, housing and education. Although the peace process created substantial improvement in the legal rights and basic protections for its Mayan population, Guatemala remains ethnically divided between indigenous people, representing between forty and sixty percent of the total population, and the nation's Ladino population, a local term used to refer to a mixed Spanish, immigrant, and indigenous heritage. Poverty and marginalization are substantially worse among the Mayan, over seventy-one percent of whom live in poverty and over twenty-one percent live in extreme poverty.

Don Mike explained that while racism, inequality, and marginalization have produced enormous suffering for the Guatemalan people, these conditions alone did not create the conflict. Instead, as the commission also concluded, structural inequality maintained through authoritarian rule and escalating levels of state repression in response to movements for social change where the root cause. "When groups lost their voice, they had no other option and lashed out," Don Mike said.

After World War II, there was cause for optimism. A broad coalition of Guatemalans rose up and overthrew the military dictatorship of General Jorge Ubico who had come to power in 1930. The coup demanded a democratic election which was won by Juan Jose Arevalo, who initiated a ten-year period known as the "democratic spring." It was a time of great hope within the country as his government set up a social security system, a minimum wage and new labor laws providing workers with the right to organize unions. Many of the reforms were modeled by the New Deal legislation of the Franklin D. Roosevelt administration.

In 1950, the optimism continued as Jacobo Arbenz succeeded Arevalo as president in a landslide free election victory. He advanced the progressive tradition of the previous government and passed an agrarian reform plan designed to address the country's social and economic inequality. The agrarian reform targeted a relatively small group of plantation owners – including the United Fruit Company, the nation's largest landowner – that controlled the vast majority of the productive land. The law did not allow the state to confiscate large plantations, but rather required plantation owners to sell to the government portions of their land not under cultivation in exchange for bonds at the price owners used for calculating taxes. The land was then to be distributed to individual rural workers and cooperatives.

But the United Fruit Company appealed to the U.S. Government for assistance to oppose the "communist" tendencies of the Arbenz regime. Don Mike explained that this was the Cold War era and the Eisenhower administration feared Arbenz policies and the growing influence of Soviet ideas in the region. In response, the Central Intelligence Agency engineered a coup in 1954 that overthrew Arbenz and ended the "democratic spring." The coup installed a military regime that replace the Arbenz government and it arrested over 15,000 people and forced thousands to flee the country.

To this day, many Guatemalans view the "democratic spring" as a symbol of a lost alternate history for the country, one in which a series of popularly elected governments might have, over time, addressed the country's profound inequalities through rational policy reform so that "The Violence" would never have occurred.

Unfortunately, Guatemala was set on a different path. One that led to decades of military rule and the systematic use of repression to protect the interests of the elite. Several times during the 1960s and 1970s, social movements led by students, the Catholic Church, unions and community groups sought alternatives to the dominant socioeconomic structure. These efforts were met by the state with threats, intimidation, attacks, and assassinations, making it difficult, and ultimately impossible, for these groups to continue their work. Death squads were formed during this time period to target the leaders of the repression. Guatemala was the first nation in Latin America where the verb "to disappear" took on a new, brutal meaning: illegal detentions, torture, execution and a process of hiding victims' bodies in clandestine cemeteries.

Frustration continued to build and inevitably, the opposition groups turned toward armed insurrection. The first stage of the insurgency began in the early 1960s and operated largely in the capital and in eastern Guatemala. The army, with the assistance of the United States, responded with brutal repression that relied heavily on a violent and increasingly intrusive intelligence service, which ultimately crushed the movement. The next state of the insurgency grew over time and peaked from the late 1970s through the early 1980s. In response, the state reaction led to the period of the most intense and sustained abuses. It was a period of time that would stain Guatemala forever.

The UN truth commission determined that more than 200,000 Guatemalans were killed in the conflict, the vast majority of whom were Mayan civilians unaffiliated with either the military or the guerrillas. They were simply caught in the middle with no place to go. State repression was so severe and overwhelming that the commission determined it met the legal definition of genocide. Don Mike had met many of the military leaders of this period and could not help but agree with the genocide definition. "Many did not view the Mayan as human beings," Don Mike said shaking his head. He also had met some military leaders who were good men who did their best to restore peace during a terrible war.

The systematic nature of "The Violence," its cruel and brutal management of power, is difficult to describe and overwhelming by almost any measure. During the most violent period in the conflict, the army and paramilitary groups committed hundreds of massacres, destroying thousands of villages, and forcing 500,000 to 1.5 million Guatemalans to flee their homes. Torture, rape, mutilation and brutal punishments were commonplace. Many rural areas were heavily militarized through policies of surveillance, forced resettlement, and the mandatory participation of all rural men in groups known as Civil Patrols.

In any conflict or civil war, all sides bear responsibility for violence and destruction, and the commission concluded that both sides committed serious violations. Nevertheless, in the case of the Guatemalan conflict, the army and state institutions committed the vast majority of the serious violations. The commission found that the state and institutions under its control were responsible for ninety-three percent of the recorded violations, including virtually all of the massacres and most cases of torture, rape, disappearance and killing. The commission concluded that the guerillas committed three percent of the recorded violations, including killing, kidnapping, and various forms of abuse. The commission could not identify who committed the other four percent of the recorded violations.

Two military regimes controlled the government during this time. From 1978 to 1981, General Romeo Lucas Garcia led the country until he was overthrown in a coup led by General Efrain Rios Montt, who controlled Guatemala from 1981 to 1983. The army developed a policy of "scorched earth" tactics that destroyed thousands of rural villages within a broad policy of militarizing the nation, especially the Highlands region. The army established bases throughout the country and forcibly recruited tens of thousands of indigenous men into the armed forces, subjecting them to brutal forms of training. Don Mike had experienced these activities first hand as he worked to restore potable water and schools in the Ixil region of the Highlands.

Don Mike explained that the military sought to "drain the water from the fish" by targeting anyone believed to support the insurgency through indiscriminate killing and mass repression. The commission documented 626 massacres, and they acknowledged that there were likely many more. These coordinated attacks commonly followed a set pattern in which soldiers and members of the civil patrol would surround a village that was suspected of providing support to the guerillas. They then would abuse, rape, torture, and kill the entire population, through spectacles of violence that sometimes lasted days to send a clear message to anyone who might dare sell food to the guerillas. In this way, "the Violence" transformed rural Guatemala. As the residents of entire communities were killed, their homes and fields burned, other communities fled into the mountains. Many became internally displaced persons and lived in the countryside for months and even years. Others became refugees, fleeing Guatemala for other countries, with around 150,000 crossing into Mexico. Some refugees continued farther north, beginning a pattern that led, over the decades, to the migration of hundreds of thousands of Guatemalans to the United States.

Don Mike shook his head sadly as he recalled the military repression that came to define the state's response as thousands of Guatemalans were arrested in their homes, at checkpoints along roads, on market day in main towns and during patrols. Many were detained on the basis of reports by spies, some of whom used their power to settle personal disputes. Others were selected based on statements from torture victims. Those detained were commonly beaten and abused and then brought to military bases, where they were interrogated by intelligence agents, commonly tortured and often killed. The commission determined that over 50,000 people were disappeared, their whereabouts unknown and their corpses buried in clandestine cemeteries throughout the country. Sometimes even now, an unknown mass grave is exposed after a landslide during the rainy season. In most cases, the grave is simply covered up as the people try to move beyond the painful memories and look forward to better days.

In an effort to undermine support for the guerrillas and to control the population, the military forced all men living in rural Guatemala to participate in the civil patrols. Its members were required to engage in continual patrols in and around their communities. At their height, the army estimated that there were over one million members of the civil patrol, which the commission found represented almost half of the adult Guatemalan men. Some patrols were unarmed and tasked with manning roadside guardhouses or overseeing their communities. Others participated directly in killings, torture, rape, and massacres against their neighbors and residents of nearby communities. The commission determined that eighteen percent of serious violations were committed by the civil patrols as directed by the army, representing a key mechanism through which the civilian population was forced to become complicit in state terror.

One day, as we bumped along a Highland's road, Don Mike relayed the story of a community that had requested a new bridge. When Don Mike visited the community he noticed a bridge crossing the same river only a mile or so upstream. Clearly, a roadway between the two sites was a much better solution than building another bridge. The community disagreed and kept pushing for their own bridge, which frustrated Don Mike as he thought they only wanted to "keep up with the Joneses" by also having their own bridge.

It was only after extensive reading, that he discovered that the community with the bridge had been forced to use its civil patrol to attack and kill many members of their neighboring community that was now requesting the new bridge. When he mentioned this at the next meeting, the people confirmed the incident, saying they had not mentioned it as they were trying to forget the terrible past. "Of course, you would not want to travel through a neighboring community that had killed your family." Don Mike said staring ahead at the road. "Although it may not be openly discussed, the past is part of the context of every project." Of course the second bridge was built after Don Mike learned this information.

Through scorched earth policies that destroyed villages, makeshift homes, crops and ongoing army and civil patrols, the state made survival difficult for internally displaced populations. In a strategy that combined the ongoing repression with a series of amnesties, the army convinced thousands of Guatemalans to come down from the mountains and live under military control. They were processed, subjected to "re-education" programs, forced to join the civil patrols and resettled in model villages and other communities under the constant surveillance of the army and the supervision of military intelligence and its network of informers.

The military's policies linked formal mechanisms of control, including restrictions on movement, with various forms of social assistance. These programs included food aid, employment and development policies, some of which were financed by the U.S. Government and various international assistance agencies. In this way, social relations among the nation's rural, largely indigenous, populations were radically transformed through constant surveillance, forced complicity in violations, and a domineering ideology of silence and submission. Don Mike recalled communities that were tight knit before the violence, having their social fabric torn apart through the mistrust and suspicion neighbors now had for each other.

Some never returned to their homes after The Violence and have chosen to remain in these "model villages". Even though the villages consist of random families tossed together into a settlement that have little or no sense of community, many simply cannot bring themselves to return to their homelands where the terrible memories began.

In 1982, the country's four guerrilla movements joined together to create the URNG. These groups often had substantial local support and hoped that, by joining together, they could overthrow the government. The Guatemalan insurgency was inspired by the Nicaraguan revolution and the military success of the neighboring Salvadoran guerillas. They imagined that they could gain large areas of the country and even planned a national strategy for controlling key regions and then marching on the capital to overthrow the government. The commission concluded that the guerillas never controlled enough combatants, weapons, or territory to militarily challenge the Guatemalan state.

Under the Carter administration, in response to the country's terrible human rights record, the U.S. Government stopped providing direct assistance to the Guatemalan military. The suspension continued through the worst years of "The Violence." As a result, the U.S. support for the country's armed forces was provided covertly and through the assistance of key allies such as Israel. Declassified material reviewed by the commission has revealed that the U.S. Government was aware of widespread massacres, torture, and other atrocities and generally supported the country's counterinsurgency efforts. Nevertheless, the full extent of the U.S. involvement in "The Violence" is still not known.

In 1983, Rios Montt was overthrown in a military coup led by General Oscar Mejia Victores. While the government continued the country's brutally repressive policies, it also led Guatemala to a democratic transition in 1985. This process involved drafting a new constitution and ending formal military rule in response to domestic calls for change, a devastated economy, and substantial international pressure. Don Mike remembered the hope during this period that the violence was finally coming to an end. Elections were held and, in 1986, Vinicio Cerezo became the country's first popularly elected civilian president in decades. Nevertheless, the military remained in control of the nation and the repression actions of military commissioner and the civil patrol continued to operate alongside ongoing surveillance and systematic human rights violations. The political transition in Guatemala was gradual. It took more than a decade to negotiate a formal end to the conflict, reduce the influence of the military, and establish the foundation, however fragile, of a more substantive democracy.

On December 29, 1996, peace between the Guatemalan government and the URNG was formally signed at a public ceremony in the main plaza in front of the Presidential Palace. The accords provided for significant expansions of education, health, and social security, an increase in tax revenues, shifts in access to land, the reconfiguration of security services and institutional support for democracy and basic rule of law principles. Interestingly, the Guatemala envisioned by the peace accords was similar to the policy vision of the "democratic spring" and embodied many of the progressive goals of social movements from the 1960s onward.

La Garrucha's Story

Sadly, the people of La Garrucha were not exempt from the terror gripping the Highlands people. As with many in the region, the people of La Garrucha were caught in the middle between the conflict of the guerillas and the army. It was a no-win situation with serious consequences no matter what choices a person made. According to Rolando, one of the community's leaders, many did not even understand the conflict as they were focused on the subsistence farming that had been their way of life for generations – they wished to live in peace and focus on feeding their families. But the events of The Violence would not leave them alone and they were swept up into the same spiral that so many of those in the Highlands found themselves in.

As the guerillas gained in numbers, they came to the farms of La Garrucha seeking corn and beans. If the farmers denied the request to sell food to the guerillas, they risked the wrath of the guerillas, including the burning of their homes, fields and in the end, forcibly taking the food. But, if they sold the guerillas the food they sought, they ran the risk of being identified as a sympathizer of the guerilla movement and could suffer the wrath of the army. Try as they might to remain neutral, the farmers could not find an acceptable way out of the dilemma.

This resulted in several incidents in 1980 and 1981 where families who were caught in the middle were massacred by the army, as memorialized in the community plaque at the entrance to the Catholic Church. The plaque lists twenty-two names of those who were massacred, including children as young as eight years old and grandparents as old as seventy. An additional twenty-five names are listed as "Disappeared," an act where the individuals were taken by the army, tortured and killed with their bodies buried in unmarked graves and never released to the families.

The plaque listing the names of those massacred or disappeared outside the La Garrucha Catholic Church

In 1981, the army defined the area as occupied by the guerillas and demanded that the people relocate to "model villages" located in San Jose Poaquil and Chimaltenango. Any persons seen from the helicopter gunships would be assumed to be guerillas and shot on sight. As part of the "scorched earth policy," the community's infrastructure was destroyed, including the school and homes. Fields were burned. Nothing was left behind that could be used by the guerillas.

But not everyone chose to move, as they feared the army and did not trust any of its directives. Hence, many retreated to the mountains to live a life as internally displaced persons, surviving off the land the best they could.

Some families even retreated to the nearby caves in the area to hide where they could find shelter and reliable water. As Rolando showed me the caves, he explained that their residents continued to farm small fields of corn and beans under the cover of darkness, carefully sneaking out each night to plant and attend to the crop. These fields were kept very small and hidden under the trees as any obvious efforts at agriculture seen by the helicopters would be burned by the army immediately. Although the families had originally hoped that hiding in the caves would be temporary, they lived in this constant state of terror for three years fearing that their position may one day be discovered. Time dragged on as the people hoped that the nightmare would come to an end and they could live their lives in peace.

In 1985, the army allowed the families to move back to the community under the watchful eye of the newly formed civil patrol. The civil patrol was made up of community members and was armed by the army. They were given strict instructions to report any guerilla activity and suspicious behavior within the community. The people left the caves, and many left the model villages to return to their homelands where they could openly farm again, but the memories were raw, and the fear continued as they never knew when the army policies may change again.

The 1996 Peace Accord was met with relief by the community members as they sought to put the memories of the violence behind them and move forward with their lives. As one member of the community said, "What else could we do? We had to put the past behind us and choose to look forward."

But the wounds are deep and slow to heal and the mistrust between the people and its government continues to this day. For some of the families, the memories are so strong that they have never returned to their homelands.

One of the areas that Don Mike and I did not agree on was religion. Don Mike was an atheist and I am a Christian. We had learned to avoid any discussions about religion. I always felt that the horrors that he witnessed during The Violence clouded his views on the topic. "How could any God allow The Violence to happen to any people?" he said. Thinking of the horrors he had witnessed, I had no response to him and quickly changed the topic.

Don Mike viewed each project as an opportunity to mend the community's fabric and restore trust between neighbors, communities and their government. He viewed this work to be more important than the project itself as it helped develop the path to sustained peace and a healed community.

"The future of Guatemala lies in the hands of its people." Don Mike said. "The restoration of a community's social fabric and its engagement in a participatory democratic government is its only hope and we need to do all we can to help that happen."

Don Mike discussing the project with a community committee

CHAPTER FOUR
The Importance of a Solicitation

"The desire for any change in a community must come from within the community itself. It takes patience to make sure that the right project is being done for the right people at the right time."
Michael Shawcross

Don Mike would never seek out projects or advertise his willingness to help. "Who am I to tell a community what they need?" he would say with a shrug to me as we bumped along on our long rides in his Landcruiser.

"The desire for any change in a community must come from within the community itself first. We, from the outside, may think we see a hardship or problem that needs our help, but if the community does not initiate the change, it is likely the intervention will fail to be sustainable over time."

He was also careful that any solicitation he received was directly from the community. Many times, another organization or a well-meaning traveler would approach him with a project for a community they had visited. He would always politely return the request back to them and ask that they have the community contact him directly. He wanted to make sure that the change being sought was being driven by the community and not imposed from the outside, no matter how well-intentioned. "The road to failure is paved with good intentions," was a saying Don Mike repeated often.

Over the years, I have since been asked to assess and resolve issues with many failed projects done around the world by other organizations. In many of these cases, the projects were identified and scoped by the well-meaning outsiders and not the community. Sadly, the lack of endorsement by the entire community subsequently led to an unwillingness to sustain the improvement, and it fell into disrepair and became inoperable.

The lack of community ownership in the project was well summed up by one village leader when he said, "We were waiting for the NGO to return and maintain their project." Once a community no longer accepts the responsibility and ownership of the project, the sustainability of the system is doomed.

With time, I came to realize that Don Mike's reputation of telling the truth and treating everyone with respect had spread across the highland region. As Don Mike waded through the many solicitations that came to him, he carefully screened out requests for assistance that might benefit only a few families within the community.

"Every project must be a community asset in the end that will be owned, operated, and maintained by the community itself," Don Mike said. "Requests that disproportionately benefit only a few families might lead to division within the community and disagreement – leading to many unintended consequences."

A few years later, another organization asked me to help assess what had gone wrong with a project and how they might get it back on track. I soon learned that the organization had only listened to a few of the well-educated members of the community because they spoke English and the organization had no translators. The voices of the rest of the community members were simply not heard. It was not surprising to me that this created friction within the community with many members believing that the project was only done to benefit the English speakers. Sadly, the project had deepened the divisions within the community and the project had even been damaged through sabotage due to the hard feelings.

Don Mike was particularly concerned that the community's most vulnerable members were not excluded from the project's benefits. Many times, especially with water projects, this proved to be very difficult. But Don Mike always refused to let the most vulnerable members of a community be left behind regardless of their economic status.

Requests for only financial assistance were also rejected. "It's not about the project, but the process of planning, building, and maintaining the project as a community that provides the value," he would say.

He saw the benefits in mending the community fabric that had been torn by the war. Using a project to, once again, pull together a community to work towards a common cause was maybe the most important part of the program. He knew that a community that invested heavily in the program would take pride in it and do what was necessary to sustain it over time. In the end, it had to be their project and their achievement. He also knew that once a community worked together to deliver a successful project, it could lead to them considering other opportunities – opportunities that they would succeed in if they continued to work together.

Don Mike said, "The future of Guatemala depends on the mending of the fabric between the local government and the people themselves. The long-term sustainability of any program needs to rely on the Guatemalans themselves and not those from the outside."

Given the history of The Violence, this was a tall order as people continued to lack trust in their elected officials, while the local government lacked a will to represent its rural constituency. Nevertheless, he insisted that every project include the municipality as part of the solicitation and project process, knowing that it would likely complicate every step along the way.

I have observed that most non-government organizations prefer to work only with the communities themselves and exclude the municipal government. This certainly streamlines the project process and simplifies the decision-making needed, but Don Mike insisted on taking the path less traveled because it held important benefits in the end.

The process of developing trust between the people and their government was an important part of sustaining the peace that had been so long in coming to the Highlands. "Yes, it will take more time and hard work – but work that is necessary and entirely worthwhile," Don Mike would say.

On a few rare occasions, I witnessed the municipality trying to defer its commitments to the project. If Don Mike was unable to sway this decision, he would let it be known that he would never work in that municipality again until the commitment was met or new leadership was elected. It did not take long for his reputation to spread across the municipalities of the Highlands, and soon everyone knew that a project promise was not to be made lightly.

Like all people, Don Mike had plenty of flaws. One of them was that he lacked the ability to forgive or acknowledge that a person could change over time. He always believed that people wanted to do the right thing – until they crossed him. Once that occurred, he refused to work with them. Forever.

Only half of all solicitations would warrant a site visit, which always included a meeting with the community leaders, the municipality, and a general assembly of the people. This ensured that everyone had a voice, and transparency was a key part of the project's decisions. Don Mike was a man of infinite patience and insisted that no project would proceed until a community agreement was reached. It did not matter if funding was only available now or a donor was insistent upon completing a project on a predetermined timeline.

"We will not start the project until it is completely ready. I think we need to give the community some more time," he would frequently say.

This led to frustration by institutional donors who measured themselves on the number of beneficiaries as promised to managers located thousands of miles away on a schedule that fit them, not the beneficiaries. Don Mike would not allow himself to be pressured by such demands and would not allow projects to proceed until they were fully ready, even if it jeopardized access to financial support.

Over the years, I observed that some funders were not able to adjust to his style and elected to take their support elsewhere. While others came to appreciate that in the end, it was not about the numbers achieved on their immediate schedule, but the quality of the projects and their sustainability over the long term.

"I wish donors would measure their effectiveness based upon the number of people who continue to receive water each and every year from their projects and not based upon the number of ribbon cuttings done for each annual report," he said. It is a lesson that I carry with me to this day.

I also noticed that Don Mike never "declined" a project that needed more time and discussion. He would always tell the community the project was "not ready yet" until they reached a consensus. This sometimes took months or even years. In one instance, the project implementation did not start until a full ten years after the solicitation was received as he waited for the community to come into agreement. Don Mike taught me that long-term planning of projects takes patience and flexibility.

I observed that Don Mike frequently used a "final test" as "homework" that he gave to a community. This is now part of the EWB Guatemala process. His homework was usually centered around the school as he wanted to see if the community would rally around a community asset and work together to sustain it. A typical assignment might be repairing the broken windows to a classroom, fixing the stairs that had fallen into disrepair, or improving the site drainage so the soccer field would no longer be muddy. If the community did not have a school building yet, it might be a small storage shed to hold the school's supplies.

Don Mike did not tell the community that they were being tested, as he wanted the homework to be done out of their commitment to community assets and not to achieve approval for a project. If the community was unable to complete such a simple project, he would not consider them ready, or investment worthy, for a larger community program.

In my estimation, maybe ten percent of the project solicitations received were deemed ready to move forward. Then the gut-wrenching process of prioritization was needed to determine which ones would be selected for funding.

In the end, Don Mike did not care if he was part of the project, he only cared that the people got help. He was always reviewing what options were available from other entities, both government and non-government, to deliver the project. This generally resulted in Don Mike working on the projects that were in the most out-of-the way locations where others feared to tread and far from a soft bed. Unlike many, he relished the rural environment, basking in its beauty during his long walks with his bedroll tucked into his pack. Sometimes he would call me and say, "Michael, I need a 'walk about.' I am going to walk across the Chuchumatanes range, so you won't hear from me for a month or so."

In my experience, Don Mike's tried and proven method delivered success in more than ninety percent of the projects that he assisted with over four decades, as they are still in operation due to the solid foundation he worked so hard to build. If the development world had walked in the footsteps so aptly left by Don Mike decades ago, untold amounts of aid money would not have been wasted on failed projects, and countless beneficiaries would be enjoying sustainable solutions to their self-identified challenges.

An Idea of a Bridge Sparks a Solicitation

As the story goes, one day, a man on a motorcycle rode down the path from the north. He crossed the footbridge over the Rio Motagua and decided to rest on the bank of the river. The morning sun felt good as it warmed him, and he began to drift into sleep. Soon he was awakened by Rolando, who was at the time a young community leader and had come to meet him. Rolando greeted the stranger with a warm smile and friendliness that made him feel at ease and welcome.

Since, at the time, the area lacked electricity and cell phone coverage was nonexistent, any news that could be extracted from a traveler was a precious gift. Rolando would not pass up this opportunity to talk with the stranger about the news and share it with his community.

"It really is an understood obligation," Rolando explained. "But I also really enjoy hearing and sharing the news and stories."

The stranger then told him about the excitement in another village, Xepanil, where a new vehicle bridge and roadway had been built. The stranger went on to explain that the bridge had been done in a three-way partnership between the community, the municipality and a group of foreigners led by a man with a long white beard, a funny hat and a strange accent. "You should see if they can help you guys out with a new bridge and road here," the stranger suggested.

As Rolando remembered the story, he told me he stared back at the stranger. Even as a little boy, he remembered the leaders talking about the need of a bridge and roadway that would provide access for the community. Over the years, they had tried several times to get such a project started – but the fact that the bridge crossed the boundary of two municipalities which fell in two separate political departments always posed hurdles that seemed impossible to clear – and of course, the bridge would need to be larger than anything that they had ever seen before.

But Rolando is a man of vision. As he watched the stranger ride out of sight, he allowed himself to dream about what changes a bridge and roadway might provide to his community. It had been years since the elementary school had reopened after the war, but it was impossible for the children to further their studies beyond the primary grades.

He, like every father on the planet, said. "I just want a better life for my sons and daughters than I have, and education is the answer."

He also knew that a roadway and bridge would allow the community to access the market in the city, easing the task of buying and selling goods. At this time, it took two days to go to the market and since the goods had to be carried back home, only the most precious of items were bought.

"I had seen the fresh vegetables in the market and talked with the farmers," Rolando said. "Some of the farmers told me how they were able to increase their income several fold by growing vegetables on the very same ground that they had once grown only corn and beans." With reliable access, his community could also sell tomatoes, broccoli and cauliflower and not have to worry about them spoiling before they could reach the market.

But the most important item for him was access to a health care facility. His mother, Gavina, is one of the community's trusted midwives shouldering the responsibility of tending to the community's health.

For as long as he could remember, he would hear his mother complain about how she could not do her job without access to proper medical facilities. She knew when people were in trouble and needed more assistance than what she could provide, but she simply did not have the means of getting them to the facilities they so desperately needed. As she watched the members of her community suffer needlessly, she was more and more frustrated.

Rolando began talking about the idea of a roadway and bridge that would connect the municipal centers of San Jose Poaquil and Joyabaj. Such a project had never been seriously attempted in this area and seemed simply impossible – but if it could be done, what amazing opportunities would be available to them.

"A bridge would remove 'the barrier' that had prevented us from moving forward after The Violence," he said. "It was the barrier to our dreams."

As discussions continued, Rolando was asked if he might lead a delegation to find this foreigner and see if they may gain his assistance.

"I knew that leading the project would put my reputation at risk. If the project failed, would anyone ever listen to me again?" But Rolando is an optimistic person who never was intimidated by a challenge – and told me that the hope he saw in his mother's eyes was something he had to pursue.

So, under Rolando's leadership, a group of delegates was chosen from the surrounding communities to seek out and meet with this foreigner. The stranger on the motorcycle had said that the foreigner resided in Antigua, a city that was two days travel away – a city that none of the delegates had even seen before. But the trip needed to be done, so the delegates kissed their families goodbye and began their journey. As almost an afterthought, one had the idea of bringing a turkey to be offered as a gift to the foreigner as they simply could not think of anything else to bring to him.

The delegation traveled to Antigua by bus. Rolando told me. "We were amazed by the elegant architecture and cobblestone streets. Everything seemed so modern and shiny."

They began asking around for a foreigner with a strange accent who built bridges in the Highlands. Soon, they were pointed in the direction of a restaurant and told them to ask for a man named Don Mike. They entered the restaurant and soon a waitress with a big smile asked. "Can I help you?" She spoke in their native language which immediately put them at ease. They asked her if she might know a man known as "Don Mike." She smiled and pointed to a table where a foreigner pored over the day's newspaper, but also took time to gaze out the window at Volcano Agua.

Rolando made the introductions. He asked if the man might be the person who helped the community named Xepanil build a bridge and roadway. Don Mike licked the eggs from his whiskers, a necessary habit he had developed over the years. He then motioned for the delegation to sit down and asked his favorite waitress to find some coffee for the travelers.

Hence, the first meeting occurred where Don Mike explained how he worked and the steps that would be needed. In typical fashion, he outlined that first, he would need a solicitation signed by the communities and the two municipalities. He asked that the solicitation include as many signatures as possible. He explained that if a person could not write, the custom of showing their support in the form of a thumbprint was perfectly acceptable. He then gave them his phone number and asked that they call him when they were ready, and he would visit the site. As was always the case during the first meeting, the only thing he could promise was that he would listen to them and give the project an honest assessment.

As the meeting wrapped up, the turkey was brought out from under a coat and offered to Don Mike, who gladly accepted the gift as they shook hands and parted. Later, Don Mike would admit to giving the turkey to the waitress, as being a bachelor, he had no idea on how to cook such a bird.

One of many pages of thumbprints from the La Garrucha Solicitation

The First Site Visit

A few weeks later, Rolando made his way to San Jose Poaquil to find a phone and call Don Mike. The community was ready and hoped that he might visit the site. A date was selected, and Rolando remembered returning to the community with a spring in his step and hope in his heart. He told me that he liked this foreigner from the very beginning as he put him at ease with his clear and simple words and an obvious respect for his people.

Don Mike arrived in his Landcruiser that was still young at only twenty years of age. Rolando led the way down the mountain path to the bridge site and Don Mike told me, "Even after spending so much time in the Highlands, I could not help but stop and marvel at the valley's striking beauty. It is stunning, even by Guatemala standards."

They rounded the last curve and saw a gathering of several hundred people waiting patiently for them. Don Mike could see from their dress that the group included both Quiche and Katchequel Mayans. The energy was high as the women fussed over a large pot of chicken soup and the men sat and discussed the project that they hoped would come to fruition.

"I was impressed as I could see that the groups were already working together," Don Mike said.

Rolando told me that he was impressed that Don Mike then graciously shook everyone's hand with a pleasant greeting – a process that took the better part of an hour. He then walked across the old, shaky pedestrian bridge to a spot on the river's bank under a large amate tree, which carries the name *chupo* in Mayan. Don Mike always loved amate trees as they were used by the Mayans as the parchment to carry their writings. When the Spanish arrived, they tried to eradicate the tree for fear that the Mayan writings would continue. Tried as they might, they were not able to remove all the amate trees and a few large, magnificent survivors were cherished by the people. As "a book man", Don Mike always had a special place in his heart for the amate tree.

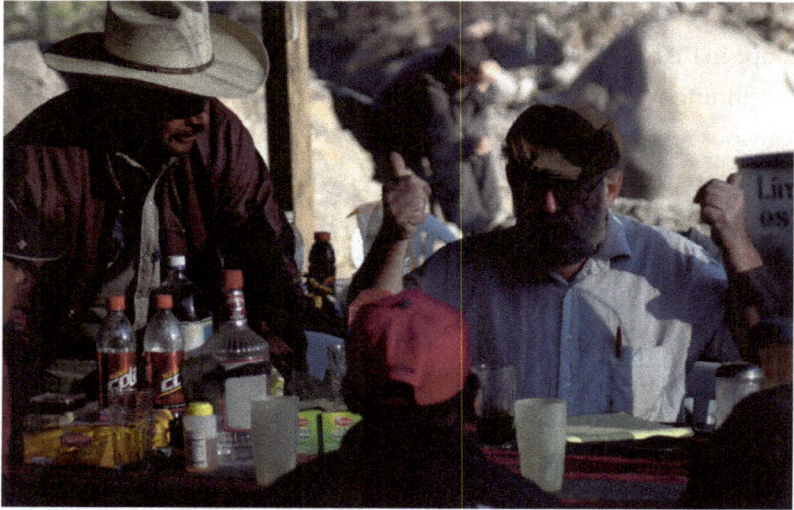

Don Mike explaining the project process to the
La Garrucha Bridge Committee

As was the custom, the meal would be served before the meeting. The lunch consisted of fresh blue corn tortillas usually reserved only for special occasions and chicken soup with large pieces of squash, carrots, tomatoes, onions and potatoes that filled the bowl till it spilled over its edge. As with any special guest, Don Mike was also offered a glass of Coca Cola, which is roughly revered in the same manner as champagne is in the United States. Don Mike told me he never liked bottled soda but knew that accepting this special gift was not optional.

As the meeting began, he knew it was important to clearly

explain the project process again, waiting patiently for his words to be translated from Spanish to the two Mayan languages. He explained the three-way partnership that was needed between the municipalities, the fourteen communities and the nonprofit group.

Don Mike said, "Yes, I know the project is large. Larger than anyone of our groups can accomplish, but together we might be able to do what seems impossible."

He continued explaining that the communities would need to organize themselves under one leader and a committee that contained a representative from each community. They would need to organize the volunteer labor needed to build the project and gather the other materials that the community may be able to provide such as wood and stones. The group would also need to coordinate with the municipalities on the roadway construction needed to connect to the bridge, including right of way agreements.

He also encouraged women to be part of the committee. "If you want to get something done, put a woman in charge," he would always say to me. This was in direct conflict with the machismo society of the Mayan and it was one area that he always tried to push the traditional culture for change. He would always gently suggest to a group, "Can you add a woman to the committee? I think it is important and we would all benefit from her perspective."

Over the years, he had learned the importance of constantly pausing and asking for questions to try to draw out the community's concerns. The meeting's choppy pace continued for two hours as questions were translated from the two Mayan languages to Spanish and back again. He asked them where they thought the best location for the bridge was and asked them to clear the vegetation from the slopes so the banks could be thoroughly inspected. He would return in a few weeks with an engineer who would determine if such a bridge might be technically possible.

As always, he continuously stressed that he was not promising to work with them on the project and did not even know if such a bridge was possible. But he told me that he could see the hope in their eyes and knew that despite his words of warning, expectations were growing by the minute. He then suggested that the communities continue to talk after he left to discuss the project and develop a list of additional questions.

As Rolando and Don Mike walked back to his truck, Don Mike asked to see the school. Don Mike remembered the building generally being in good repair, but there was trash everywhere. He expressed his frustration to Rolando that a land that held such amazing beauty was tarnished by its people with trash discarded randomly. Rolando agreed and added his hope that someday the people of Guatemala would see the amazing gift that was given to them and take care of it appropriately.

As they reached the truck, Rolando remembered sheepishly clearing his throat. His mother had insisted that he also mention the communities need for clean water to the foreigner. As the community's health care worker, she knew what was needed to keep them healthy. But how could she do so without clean water to even wash their hands? Rolando knew the timing was not perfect, but he had promised his mother, so he asked Don Mike if he might also entertain a solicitation for a drinking water project.

Don Mike liked Rolando and his energy and willingness to work tirelessly to help his people. He responded that he would be happy to receive the solicitation, and who knows, maybe it would become a reality in the future. First, they would focus on the bridge project that would benefit all of the fourteen neighboring communities.

My First Site Visit

Don Mike and I returned to the site a few weeks later. At this time, I had already worked with Don Mike on the design and construction of four bridges along with several schools and water systems. I enjoyed lending my engineering expertise to the projects and felt Don Mike valued my judgement and opinion on the technical details.

"You engineers are certainly unique and bring a different perspective to a project," he said. "It's nice having you around, even though I sometimes have no idea what you are saying."

I was worried that the bridge would be an especially difficult project based upon Don Mike's description and feared that maybe it was simply not possible given the materials and equipment available. I was concerned that a difficult "Go – No Go" decision was ahead of me. I had always hated such decisions when the project was held in the balance by its technical challenges, not the community's ability to organize.

As I saw the site for the first time, I remember feeling a sense of relief. The decision would not be difficult at all – this bridge was simply not possible, I thought. The length and height were half again more than any bridge that we had attempted before. I observed that the force of the river would easily sweep away temporary supports used to do the construction, requiring a completely different design approach.

I shook my head at Don Mike and suggested, "Let's keep the discussion short and explain that this project is simply beyond our capability. We don't want to raise any expectations."

Don Mike asked me to take a closer look and maybe some advice could be offered to the group. I pulled out my tapes and tools and did my measurements and checked the rock that formed the banks on both sides of the river. I then stripped off my pants and put on my wading shoes to inspect the bank scour under the existing riverbank.

Even though no rain had fallen in over a month, the current tugged strongly at my legs. The rocky streambed was swept clear due to the fast-moving water and I could see the rocks deposited downstream where the stream widened, and the slower moving water dropped them. As I probed the bank with a long stick, I looked up and saw over a hundred people looking down at me from above. I remembered seeing the hope in their eyes as they watched me do my work. How could I ever tell them "no"?

I soon found Don Mike and wrinkled my brow. "Yes, the project is more difficult than anything we have done before Don Mike, but I promised to do my best to find a solution."

Don Mike winked at me with a twinkle in his eyes. "I knew you would be up for such a challenge," he said.

We sat under the amate tree where another meal was served with blue tortillas, chicken soup and of course, Coca-Cola. After Don Mike finished the last of his Coke, he again laid out the project process and answered questions. Don Mike was always very patient and knew the importance of repeating the same message over and over again.

The expectations were again clearly explained, and as they were laid out one by one, I could see the doubt start to set in. Fifty volunteers would be needed each day to work on the bridge. The communities would need to gather enough stones to fill a small house. Each of the requirements was enough of a challenge and when they were piled on top of each other, I could see in the expressions on the faces that they were daunting.

Don Mike also saw the doubt starting to sink in. "I know I am asking a lot from you," he said. "This project will demand the best effort of everyone – but nobody is in this alone. I have faith in you. You are descendants of a great people who built amazing temples. I know you can do it if you all work together as one team. I will be happy and honored to walk with you along this journey," he said.

As we walked back to the truck, we passed near the school and I remember Don Mike insisted that we take a detour and visit the site. We found the school grounds clear of trash and trash bins stationed at various places along the grounds. We even found that the trash was being separated into compostable and non-compostable materials. I then remembered Don Mike smiling - he now knew that he had a community partner that was ready.

An Agreement is Finalized

After many more meetings, including a town hall meeting in each of the fourteen communities and more "homework" assignments at each of the schools, the day finally came where the agreements would be set, and a schedule developed. Don Mike and I traveled to the meeting which would be held in the offices of the municipality of San Jose Poaquil. Representatives from the seven communities north of the river would be there along with the representatives of the seven communities south of the river. The municipality of Joyabaj was to be represented by the Vice-Alcalde, Florencio Gomez.

Florencio is a tall, handsome man with a smile and laugh that quickly makes friends. He carefully takes the time to listen to everyone intently and lets them know that their opinion really matters to him. At the time of the meeting, he was in his late thirties which is young to hold such a position in a society that respects and listens to its elders. He clearly was destined to have a long and successful political career.

As is the custom in Guatemala, nothing starts on time and the group slowly assembled for the important meeting. Quiet, nervous whispers filled the room as representatives from all the groups made their greetings and caught up on the daily news while sipping their bottles of Coca-Cola. Finally, the *alcaldes* (municipal mayors) arrived and the meeting was ready to begin. The pace of the meeting was slow as each message needed to be translated into both Mayan languages and then back to Spanish with any questions and replies.

The meeting started with the goal of the project being clearly articulated by Rolando. He quietly described how access was needed to schools, markets and health care facilities for all fourteen communities and that this would require a bridge and roadways that would accommodate trucks and buses.

He also added his dream that a bridge and its benefits would entice some of the families who left due to The Violence to return to their homelands. As he finished his speech, several people applauded. Then the applause was repeated twice more as the message was translated into the two Mayan languages.

I remember that smiles and bright eyes were everywhere as the group dared to believe that their dream may become reality.

Then the scope of the project was laid out by Florencio, and I could see that it seemed daunting to the attendees. The roadways leading up to the bridge from the south would need to be widened and straightened to accommodate large vehicles. This would require considerable effort as the current narrow path was notched into the mountain in several places. The municipality of San Jose Poaquil responded by pledging to hire a large backhoe to do the work.

As difficult as the roadway to the south seemed, the roadway to the north appeared to me to be nearly impossible. The portion from the top of the mountain to the river was only a crude motorcycle path and a new roadway would need to be cut into the mountain to provide a safe descent of over 3,000 feet to the river.

I cringed as I feared I might need to design a path for the roadway which avoided the large slides and sheer cliffs that existed on the mountain face. But Florencio explained that he would hire a large bulldozer to do the work and the operator was an expert in building mountain roadways in the region. He knew the work was difficult but was confident that it could be done.

Then I described the bridge that would be built and waited for it to be translated to the attendees. It would be made of concrete beams and span the entire river without any supports which might impede the flow. Of course, being an engineer, I produced sketches that were drawn on a large sheet of paper taped to the wall that showed the bridge's length and width. I explained that materials such as cement and rebar would need to be brought to the site using the improved roadways, so that work would need to be done first. The communities would need to gather enough stones the size of soccer balls to fill ten dump trucks which needed to be stacked nearby prior to construction. Finally, I suggested that the actual construction of the bridge itself would take two months and require fifty workers each day.

As I sat down, silence filled the room. All the enthusiasm at the beginning of the meeting was gone as the group realized the tremendous amount of work that lay in front of them. The group looked down at their hands in silence wondering if the project was simply too big. Could their community respond? Would the other communities and municipalities respond? Could everyone be counted on to meet their commitments?

It seemed that Florencio then realized that the project was hanging in the balance. He stepped forward and delivered a passionate speech for the ages.

"This is an opportunity of a generation and we owe it to our children and grandchildren to rise to the challenge." he said.

As his words filled the room, the translators did their job without waiting for him to stop so as to not lose the momentum. I noticed that the faces of the group began to change as people stopped looking down at their hands and raised their heads to focus on the fiery speaker.

"I know this is a lot of work, but I have faith in all of you," Florencio continued. "I know you are strong and we have many hands to help. We all want to see the barrier broken that has been holding us back from reaching our dreams. I know I can count on each and every one of you to do all you can and more, so I will begin the roadway work immediately, without hesitation or concern. The only question I have is what should we name of the bridge?"

The group looked at each other with surprise. They had not allowed themselves to think of a name for the bridge before, but now it seemed so important to do so. I think Florencio somehow knew that a name would make the project real and inspire the group.

Several names were bantered about the group and finally the name "The Friendship Bridge" was suggested. They all agreed that since so many different groups with different backgrounds would be working together on the project, it only seemed fitting and right.

Don Mike shot up out of his chair, cheering and clapping his hands. He had come to value peace above all else based upon his life's experience. His Landcruiser bumper always sported a "COEXIST" sticker and he taught me that any community's future always started with peace. The idea of a "Friendship Bridge" clearly pleased him.

And so, it was. The Friendship Bridge project was on its way from being a dream to reality.

Don Mike sampling some cusha at his home in Antigua

CHAPTER FIVE
The Students' Bridge Design

"The design must be adapted to fit the local setting, materials and tools instead of dictating them. The designers must be willing to change their approach, means and methods to work in this environment."
Michael Shawcross

Although it might have been easier to work only with experienced engineers, Don Mike saw the value of working with engineering students in the design process. "Most of the world lives in conditions similar to Guatemala or worse," he said. "It is so important that young engineers understand how to develop solutions that work for the other eighty percent of the planet." As a world traveler, he also knew the value of expanding the students' perspective by exposing them to different geography, people and cultures.

"The future of the world does not lie with you and I," he told me. "It rests with the young people and it is our obligation to share with them all that we know before we are gone."

I remember walking into the classroom at Marquette University to meet the design team for the bridge for the first time. It was a beautiful fall day and the campus was full of the energy that only a university can bring. Four bright eyed and nervous students looked back at me as I unpacked my bags and surveyed the room. This was not the first time that I had agreed to mentor a senior design team at Marquette University, and each group was always unique.

I looked forward to it each year and although I had never had a bad experience, I was also a bit anxious wondering who the design team might consist of. I loved working with the young engineers and guiding them through the design process, watching them learn and make decisions. Also, as an engineer, I felt that it was my professional obligation to help the next generation of engineers, as my mentors had done with me so many years ago. It is an important part of the engineering profession. To become a professional engineer in most states, a university graduate engineer must also work under the supervision of a professional engineer for four years before being allowed to sit for the final board exams. This "apprenticeship" relationship has guided the profession for many generations.

Prior to the meeting I had sent out a request to the civil engineering senior design class for a team to be assigned to a project in Guatemala. The project description was clear – this project would require more than double the effort that would be needed for the other projects that a team could choose. It would require an unusual amount of dedication and attention to be successful.

But this project was real. Not some made up exercise or duplicating a design that had already been built. This was a real project that would help real people.

I had learned over the years that such an approach would typically land me a design team of the best students who were motivated by more than a grade or graduation. Any students interested in such a challenge were already taking the Civil Engineers oath seriously – "To protect the health, safety and welfare of the public." They were interested in service and understood that "public" meant the planet and not only their own backyard.

As was required for the class, the team needed to present their qualifications to me and their professor, Dr. Daniel Zitomer, or "Dr. Z" as he is called. Dr. Z is a gentle man who always seems to be optimistic and sees the best in everyone and every situation. I had traveled with him to Guatemala on past student senior design projects and his "even-keel personality" was always perfect when an unwelcomed change occurred.

"We'll figure it out. Don't worry," he would always say when the students, and sometimes I, were ready to freak out.

The presentation from students selected by Dr. Z. was in an interview format similar to what a consulting engineer would make to a client for a prospective project. The students stood at the front of the classroom nervously fidgeting and chatting. The students were dressed in business suits, taking care to look their very best and professional. Although, I have to admit, a couple of the guys could have used a haircut.

I smiled thinking back so many years ago when I was interviewing for my senior design project. Did I really look so young and nervous? I greeted each of them with a handshake and a smile.

"It is so nice to meet you! How were fall midterms? Do you think the Marquette basketball team will do well this year?" I did my best to break the ice, but the awkward nervousness continued.

James had been selected by the other three members of the team to be the project manager and leader. He had an athletic build, blonde hair and a born leader. He was a native of St. Louis and had attended Marquette because of its motto—"Be the Difference."

Passionate about service, his attention was grabbed when he had heard about previous Marquette senior design teams and that they had done real projects in Guatemala. He was amazed by the idea that he, as a student, could actually design a project that would be built and help people. He had quietly listened to previous team members tell their stories and experiences, which fascinated him. Since his sophomore year, he knew he was destined to lead a team during his senior year and his moment had finally come.

He started the presentation in a professional manner, outlining the roles of each team member.

Mollie would lead the structures design. She was a strong woman with broad shoulders and a laugh that could not help but make you smile. As she gave her presentation, she clearly was a bit intimidated and nervously fidgeted with her hands as she explained her role.

Paul's role was the geotechnical and hydraulic engineering. He is of slight build and quiet demeanor. He explained that his native language was Spanish, and he would also serve as the team's translator. Although quiet in nature, he gave his presentation with confidence and conviction.

Andy would lead the transportation design along with material testing, specifications and construction. He was a large fellow and had a quiet confidence. He had lived his entire life in Milwaukee and was excited to experience other cultures while contributing his engineer skills. To help pay his way through school and gain experience, he had worked for a local engineering firm as an intern. His familiarity with how the engineering profession developed and delivered a project would prove to be a valuable asset to the team.

"Yes, we know we will have to work harder than any of the other teams," James stated looking me square in the eye. "But we realize that this is a tremendous opportunity to work on a real project that will benefit real people. We are motivated and committed to the project and the community. We are more than willing to do the extra work needed to make sure the project is a success. I know we may not have a lot of experience, but we are eager and willing to learn."

I could not help but be impressed.

Marquette University's History of Engineering Service

At our first meeting together, I told them the storied history of the Marquette University bridge program in Guatemala. The university has long had an engineering service culture that formed even before EWB-USA existed.

Inspired by its moto of "Be the Difference," hundreds of engineering students had planned, designed and constructed dozens of bridges, water systems and schools in the Highlands region of

Guatemala. I explained that the design team would be riding on the shoulders of their predecessors who had worked so hard to build the program's reputation and relationships. They would be responsible for protecting and advancing that reputation.

Marquette University was connected to Don Mike by a local civil engineer, Paul Johnson. Paul had worked around the globe assisting the most vulnerable with their infrastructure needs and had also fallen in love with the Mayan people during his first trip to the region in 1992. He then worked with Marquette University to establish a senior design program that would design and build bridges in the Highlands.

"We all have mentors throughout our lives," I said to the students. "How lucky and honored am I to have both Don Mike and Mr. Johnson as mine."

Mollie said, with a surprised look on her face. "You, even now, have mentors?"

"Yes, even us 'old engineers' still need mentoring," I replied with a smile.

The program has always been inspired by the history of the Catholic Church and bridge building and I felt it was important for the students to learn and understand this history.

Rob was one of the alumni of the Marquette program and had been the project manager of my first Marquette Senior Design Team. Rob was "an engineer's engineer" and looked the part being of slight build and wearing glasses to convey his extraordinary intelligence. Rob was clearly one of the best students I had ever had and has gone onto an amazing engineering career, tackling structures around the world.

He has always remained a staunch supporter of the Marquette

program, even when he was overseas. Every year or two, he would find a way to travel with a design team and assist them with the construction, sharing his knowledge and experience. Being a man of extreme faith, he had researched the history of bridge building and the Catholic Church and had taken the time to write down this history which was handed to the team.

Bridge Building in the Dark Ages

As the Roman Empire fell into decline, and the raiding Goths, Vandals, Huns, and Saxons battled for control of western civilization, society's attention turned from science and engineering to war and plundering. The Dark Ages saw much more in the way of fortress building than of bridge building, and, as commerce and communication decreased, the knowledge of engineering and construction cultivated by the Romans was all but lost. Many bridges were neglected and fell into disrepair or were destroyed by war.

Travel during the Dark Ages was difficult and far from safe. Travelers found themselves at the mercy not only of the harshness of the elements, but also the ruthlessness of highway bandits. Other enterprising barons exacted heavy tolls from passers-by along popular routes. Where roads met water crossings, travelers were left to either find a shallow point to ford or to test their skill at swimming across. At some traveled crossings, boatmen would ferry travelers to the opposite shore for a fee, though all too often the boatmen proved to be pirates in disguise.

Out of compassion for the weary traveler, the Church led efforts to improve safety and hospitality along pilgrim routes, particularly at river crossings. The Church worked to establish more reliable ferries and commissioned the construction of many hostels and simple bridges. The Church's primary role in the execution of such projects was the collection of funds. Donations were gathered from the faithful, who gave willingly in hope that their act of piety would bring eternal reward. The physical work of bridge construction was generally led by guilds of traveling artisans. The structures were simple and usually small enough to be built of timber.

Brother Bénoît

For hundreds of years, bridge engineering saw little in the way of advance. It was not until the 12th century that large stone bridges rivaling those built by the Romans again were attempted anywhere in Europe. Then, in 1164 in the south of France, a bridge was completed that spanned the River Durance at Bonpas. The project, which signaled the end of the Dark Ages and the beginning of a revival in the study of bridge engineering, was led by a Benedictine monk named Bénoît.

In the years that followed the completion of the bridge at Bonpas, Brother Bénoît and others who worked on the bridge, both Benedictines and lay artisans, continued to work towards providing better accommodations for travelers. Throughout the south of France, they established hostels, hospices, and ferries. Their work also included the building of bridges at Lourmarin, Mallemort, and Mirabeau. Their association remained strong, and the services of the group were continually in demand. Eventually the group came to be called the *Frères Pontifes*, or the *Brothers of the Bridge*.

The brotherhood was not strictly a priestly order. Its membership mostly consisted of people from three categories: clergy, knights, and artisans. Women were allowed. The clergy of the group generally led the commissioning of projects and organized the collection of funds.

Knights, particularly Hospitaller Knights, were among the chief financial contributors to the work of the brotherhood. Finally, skilled artisans conducted the physical work of bridge building.

The most notable work of Brother Bénoît and the *Frères Pontifes* was the bridge over the Rhône at Avignon. The bridge was a technical masterpiece. It consisted of twenty-two arches, comprising a total length of approximately 3,000 ft. Four of the arches have survived to the present day. All of the remaining arches have spans that eclipse the longest of any surviving Roman arch. In addition, the arches at Avignon were elliptical, instead of semicircular. Thus, the roadway was carried higher above the river than it would have been had semicircular arches been used.

While the technical achievements realized in the bridge at Avignon are in themselves impressive, the project is believed to have been directly inspired by a miracle enacted by a shepherd boy named Bénèzet.

Saint Bénèzet – Patron Saint of Bridge Builders

The Legend of Saint Bénèzet

In the year of our Lord 1177, Saint Bénèzet began to build the bridge as it is thus written:

On this day as the sun was setting a child of only 12 years named Bénèzet was watching his mother's sheep in the pasture. Jesus Christ revealed himself to the boy and said to him

"Leave the sheep that you are tending, for you shall build a bridge over the River Rhône. Have I not told you to have faith?"

"My Lord, How can I build a bridge over the Rhône?"

"Come with me and I shall teach you."

Entering the city of Avignon, Bénèzet went to the cathedral and found the bishop preaching to his people. He interrupted the bishop, proclaiming, "Listen to me and understand, for Jesus Christ has sent me to you that I may build a bridge over the Rhône."

The bishop, hearing these words, believed him to be possessed and in the throes of madness. He ordered his magistrate, the vicar, to skin him and cut off his hands and feet so that the Devil could not afflict his people.

Bénèzet, hearing the bishop's order, again proclaimed, "My Lord, Jesus Christ, has sent me to this city to build a bridge over the Rhône."

"You puny person, you shepherd boy," answered the vicar. "You declare that you will build a bridge where neither God, nor Saint Peter nor Paul, nor Charlemagne, nor any other could? What a wonder that would be! Wait. I know that a bridge is made of stone and lime. I will give you a stone that I have in my palace, and if you can move it and carry it, I will believe that you can build this bridge."

Bénèzet, putting his trust in the Lord, turned to the bishop and said that he would do it.

They went out from there to the bank of the river, the bishop, the vicar, and all the people gathered, and Bénèzet took by himself the stone that thirty men could not carry. He took it as lightly as if it were a pebble and placed it where the bridge would take hold.

After his miraculous founding of the bridge at Avignon, an event recognized by the Catholic Church, Bénèzet oversaw construction for seven years, until his death in 1184. Bénèzet was officially canonized a saint by the Catholic Church. His feast day is celebrated on April 14th. Bénèzet is most often portrayed as young boy with a large stone on his shoulder.

After hearing the story, the team was even more inspired and adopted the name Hermanos Del Puente (Brothers of the Bridge) as the name of their team. They realized that individuals that are inspired, have faith and are not afraid to pursue their dreams can do amazing things. James, the student team leader, who happened to be a pretty good artist in his own right, drew up a logo that also captured The Friendship Bridge and its meaning.

Hermanos Del Puente Site Visit

Similar to past senior design projects, the student design team would visit the site of their design with some financial assistance from a Marquette travel grant. The team and I would collect data on the topography and geotechnical conditions. The trip would be during Marquette University's winter break and coincide with the construction of the bridge designed the previous year. This would allow the design team to also participate in the construction of a similar bridge using local tools and materials. But, the highlight of the trip for the students was always experiencing the Guatemalan Highlands culture and meeting the people who would benefit from the project.

It was also the first time that any of the students had traveled internationally and the excitement was high. They read books about the history and culture of the Highlands. I talked with the students' parents, easing their concerns about safety and security. I always tried to do this proactively asking the students' if I could call their parents and answer any questions knowing that any parent would be concerned. I knew I would want to talk to the person who would be responsible for my son or daughter's safety instead of only relying on a written procedure.

Permission was granted and the day finally came for the trip. *Hermanos Del Puente* sat at the airport, double checking their packing list of tools, equipment and clothing. I reminded them with a smile that it was a too late to make any adjustments now, but the students continued to pull everything out of their bags to convince themselves they had everything they needed.

Upon arriving in Guatemala, the students met Don Mike for the first time. They had heard the legend of Don Mike from upperclassmen who worked on previous projects but there was nothing like meeting him in person to truly appreciate the man. I watched as they were captivated by Don Mike's stories and British accent during the roadway trip to the site, hardly taking their eyes off him to enjoy the amazing views that surrounded them.

Don Mike was in his glory and loved talking with the students. He loved sharing all he knew about the country he so loved and the people who lived in it. Each question was followed by another Don Mike story that would add to his legendary status. The conversations would continue into the wee hours of the night during the trip and Don Mike never tired of the questions or the company. The students could not help but be mesmerized by this stocky man who always wore a British flat cap and told his stories with a British accent.

The team also observed and learned the construction practices of the Highlands during the building of the bridge designed the previous year. They had heard the stories but doing the work themselves was something completely different.

"I can't believe how heavy these pails of rock and sand are," Molly exclaimed as she wiped the sweat from her brow.

The mixer was fed the buckets of sand, gravel and cement in what seemed like a never-ending dance. The process would be repeated over a thousand times before the bridge would be completed. Andy worked with the rebar, cutting and bending it to the specifications in the plans. At night he rubbed his sore shoulders from the repetitive work while calculating how to cut the twenty-foot bars into their required lengths and leave the least amount of waste material.

Carpentry was James's forte and he was fascinated by its challenges. The wood was hand cut so each board was similar, but not exactly the same. This posed a challenge in trying to find boards of just the right width or thickness for the needed purpose. Many found the work frustrating, but not James. He loved the smell of the wood and learning from the other experienced carpenters on the project.

Finally, the day came to visit their bridge site at La Garrucha and they excitedly piled into the Landcruiser. We drove the bumpy road from the north as far as we could and stopped at the top of the mountain to view the valley for the first time. I remember hearing the audible gasps from the team as they saw the valley, amazed by its beauty.

The new roadway down the mountain was under construction and they met George, the bulldozer operator. George is a stocky, middle aged man with decades of experience building mountain roads. His bulldozer is almost as if it is part of his body, as he and it work in perfect motion. He is quiet and friendly and never seems to be in a hurry.

We were amazed as he told the story of how he selected the route. A project like this might take years in the United States to complete the soil borings, surveys and geotechnical studies needed to select the roadway location.

George had a different approach. One that he had perfected by his many years of building roads in the Highlands mountains. He carefully studied the route options by sitting on the other side of the river and studying the mountain's face for several days. He then carefully walked the route and placed stakes to mark the alignment, adjusting as needed when he found hidden problems. He used his years of experience to identify the areas such as poor rock or hidden slides while also visualizing the grade of the roadway that would be needed for buses to pass.

The young engineers then realized that although George had never attended a university, he was more valuable than any engineer for this task.

Don Mike chuckled at the young engineers and their amazement as he said, "There are some real craftsmen here and it is important to remember that when you run into a problem. More times than not, they will know the solution based upon their experience. They live with these problems and solve them every day."

As we trudged down the mountain path to the river, we could see several hundred people from both sides of the river waiting for us. They were there to ensure the students would receive the best greeting possible. A large pot of vegetable soup the size of a bathtub was simmering over a wood fire and being gently stirred by a woman with a large ladle.

A group of men stood ready with shovels and picks to dig test pits that would be necessary to study the foundation. Another work team stood at the ready with their machetes to clear any brush that might be in the way of the survey team. And of course, a Coca-Cola was handed to each of the students who eagerly quenched their thirst while Don Mike choked down yet another bottle of the soda.

It was then that the students reluctantly came to Don Mike with a problem. They had left their survey equipment's battery in the charger back at camp, a half day trip away. Don Mike laughed knowing there was nothing that they could do that day to solve the problem, but a valuable lesson was learned.

The *Hermanos Del Puente* team sheepishly returned the next day to complete the topographical survey trudging back down the mountain and up again with their equipment. They complained about the soreness in their muscles from the previous day's hike and Don Mike gently reminded them that this was a trip the community made on a regular basis as they were passed by grandmothers with their grandchildren in toe.

"Can you imagine making this walk if you were pregnant or sick?" he asked, reinforcing the reality of the benefits of the bridge.

The team diligently worked to gather the data needed with assistance from the community members who worked tirelessly cutting down brush with their machetes. That evening the students slowly trudged back up the mountain and returned to the bridge construction site. They were tired but happy.

The *Hermanos Del Puente* team now focused on the bridge construction, keenly observing and participating in each step of the bridge building process. Finally, after several weeks, the neighboring bridge construction was completed, and the team saw the joy in the eyes of the community as the barrier to the community's dreams of an improved life with access to healthcare, schools and markets was realized.

On the way back to the airport, the team was strangely quiet. They processed all that they had seen and done during their stay in Guatemala and reflected on how they would be returning to a life so much different than that they had seen.

"You've got yourself another good group of engineers." Don Mike said to me as we bounced along in the Landcruiser back to the airport. "I love seeing young minds opened up to the world and its needs."

As he leaned back in his seat and pulled his British flat cap down over his eyes for a nap, he said, "The future is in good hands."

When the *Hermanos Del Puente* team returned home from the trip, the excitement could not be contained. James remembers racing to tell his friends and family about their experience and show them his photos. All the team members shared the experience of meeting the people they were going to help and their hospitality. They had looked into their eyes and knew that completing the project was not an option, no matter what the obstacles.

Now the real work began on the project and the students entered the lab to perform testing on the materials they had brought back with them. The concrete cylinders were tested to determine the mix's compressive strength. The rebar was stretched to determine its properties and even the wood was tested to determine its strength.

They had performed similar tests previously as part of their studies, but now the tests had a whole new meaning. The information would be used to inform the bridge design and ensure its safety.

I explained that a different approach was needed from the typical engineering process where the engineer specifies the material strengths, expecting the builder to find the appropriate materials to complete the project. The students had seen that it was not possible or prudent to achieve these specifications using the locally available materials and heeded Don Mike's advice to adjust their designs to what could be found locally. I could see that the learning had begun.

Several alternatives were analyzed and evaluated. They included bridges made from steel beams, concrete beams and concrete slabs. Each alternative was evaluated for cost, constructability and durability. In the end, a concrete cast-in-place beam structure was selected because it not only maximized the use of the locally available materials, but also maximized the local talent who were experienced with concrete construction.

Most bridges in the United States use steel or precast concrete beams that are made in a factory and set in place using large cranes. The cast-in-place beam bridge would require wooden forms over the river and concrete poured inside of them after the rebar had been carefully set – basically building the concrete beam in its final place.

All the materials must be readily available in Guatemala. The cement was manufactured in country and the rebar was made from recycled steel at a local plant. The sand, gravel and wood needed would all be obtained locally at the bridge site. In this manner, the project maximized the use of local workers while minimizing the cost of the bridge. It was also the type of construction that was familiar to the local workers which would maximize the use of their talents. Instead of picking up a heavy concrete beam using a crane, the concrete would be placed inside the forms - one wheelbarrow at a time.

The students pored over their calculations, taking care to have another team member check the work before handing it over for my final review. As they solved problem after problem, one challenge remained for the team. How could the wooden formwork be designed so it would hold the concrete in place until it would harden and gain strength?

The method used by previous projects of using dozens of wood columns extending from the river bottom to the bridge was not feasible at this site due to the swiftly moving water. The water would simply wash the columns away. The students queried contractors near the university and learned that in the United States, a metal beam would be placed by a large crane to solve the problem. Knowing this was not an option for their site in Guatemala, they scratched their heads searching for a solution. As I watched them struggle, I did not let them know that a solution had not yet become apparent to me either.

The problem clearly needed some additional expertise and I recruited my friend and expert bridge engineer, Kevin Hagen. Kevin had designed over 100 bridges in the United States but also had assisted with the design and construction of several Guatemalan bridges.

He is also the son of a carpenter and worked with his father every chance he could. His combined skills of bridge engineering expertise and carpentry skills were exactly what was needed to solve the problem. He recommended that a wooden truss be used to make the construction feasible.

A truss is a narrow, but deep structure made from a continuous cord or beam of wood at its top and bottom. Smaller pieces of wood are laced between the two cords to form a continuous structure that behaves as one object. It is a lightweight solution that is typically used in the construction of a home's roof.

During his trips to help build previous bridges, he knew that it was one thing to build a truss in a factory and quite another to build one on the side of the mountain using local farmers as carpenters. After considerable study that even involved building a few prototypes in his garage, he came up with the solution that utilized the local pine wood and could be fabricated with only nails and bolts – a solution that could be implemented in Guatemala.

Kevin proudly showed me the selected prototype while his wife shook her head at the giddy engineers. He explained the assembly process. Every detail had been sorted out and each step in the assembly process simplified to fit the anticipated workforce. It was what the engineering community would refer to as "An elegant solution."

The truss had never been seen before by the engineering world and to this day is referred to as the "The Hagen Truss." The students were fascinated with the solution and pored over Kevin's calculations and drawings. They replicated his design steps, learning how to use simple, locally available materials such as nails and bolts assembled in detailed patterns to make the complex connections at each joint. They learned that sometimes the most creative solutions do not involve modern materials.

As the design neared its completion, the team readied itself for the annual Marquette University College of Engineering design competition. Each senior design team across the College of Engineering needed to prepare a poster describing their project. The projects include all the disciplines of engineering and ran the gamut from electric cars, solar powered machines, engineered artificial limbs and civil engineering projects. The competition would be stiff for the top prize, but I made it clear that a grade of "A" would not be possible if the team did not win.

Hermanos Del Puente took up the challenge. They had a great project with a great story, but also needed to brainstorm for a unique, award-winning idea. They needed something that would stand out in the judges' eyes.

The winning idea was a series of photo renderings that showed the construction progress each day over laid on site photos using the unique construction methods in Guatemala. Years later, such a presentation would become more common, but I had never seen anything like it at the time. The idea carried the day and the *Hermanos Del Puente* design team was awarded first place by the college of engineering – and they got their "A."

Don Mike with community members
reviewing the bridge crossing location

CHAPTER SIX
The Construction Begins with a Surprise

"This is Guatemala. Be prepared for anything."
Michael Shawcross

As we began the bridge excavation, I looked at the darkening skies to the north and frowned. It was October 2005, during the heart of the Guatemalan rainy season, but the dark skies were different than the typical daily thunderstorms. These were dark, ugly and boiling.

It had been raining hard for days and the rains gave no indication of ending soon. There was something strange about the wind as well. It was not strong, but somehow seemed restless – almost like it was being held back and waiting to unleash its fury. I knew that this would be trouble for the project as the rain pelted my face and dripped off my wide brimmed hardhat.

The other team members consisted of about a dozen workers from the community, Don Mike, Mincho, our construction master mason, and Adam from PAVA, a Guatemalan non-profit organization.

Adam was a Peace Corps volunteer assigned to PAVA and was a perfect match for the work. He was raised in Costa Rica and his Spanish is perfect. In fact, he is more comfortable in the Spanish culture than that in his later hometown of St. Louis. After getting his degree in Geographic Information Systems, he worked for a few years honing his skills before he and his wife, Katie, decided to join the Peace Corps. He has a quiet, patient disposition with a welcoming face and smile that makes it easy for him to make friends. His technical background made it easy for him to learn the engineering tasks necessary for his job and his work ethic quickly endeared him to me. We had become fast friends. Unlike some other Peace Corps Volunteers, he is focused and knows why he is in Guatemala. Simply put, he wants to help people.

Katie, his wife, had accepted maybe the most difficult of all Peace Corps assignments. She is a tall woman with dark hair and dancing eyes that are magnified when she flashes her wide smile. She worked with the junior high school students teaching them sexual education – a difficult task in any environment that is especially challenging in the conservative culture of Guatemala. Katie and I have become good friends as well, and I typically stayed at Adam and Katie's home during my frequent trips to the country. Of course, I would bring rare treats from the US such as M&Ms and Mountain Dew which we would enjoy late into the night over our long conversations about the St. Louis Cardinals baseball team and other news from the States.

Mincho is an experienced mason who has worked with PAVA on water and school projects for many years. He lives in the Highlands and is an expert in the construction methods of the local builders. Simply put – he is a builder and loves his trade. But he is not only a builder of infrastructure, but also of communities. He is a middle-aged man of an average height and build – but that is the only thing that is average about him. At that time, he already had over twenty years of construction experience, having started as an apprentice as soon as he could swing a hammer. He always knew that building things was his calling and by this time, he had completed dozens of water systems, schools and homes.

But his greatest talent might be motivating and building communities. His work always required him to labor side by side with the community's volunteers while keeping them motivated. This can be especially challenging during long projects that require months of physical work. He possesses a welcoming smile and friendly personality that people are drawn to. Famous for singing while he works, he always can keep the workers spirits high and helps pass the time during the back-breaking labor.

I had learned that somehow he always seems to know what is needed to keep the people engaged in the project. With an uncanny sixth sense, he seems to know what is needed to pick up the community's spirit at just the right time to keep them engaged. I remember one school project for a small community where the workers were clearly out of gas. They had been working for months and fewer and fewer volunteers were showing up each day.

I was frustrated and angry and even suggested pulling off the project for awhile when Mincho told me to "Take it easy." He announced the next day there would be no work and we would all play soccer instead! I was furious that we would waste a whole day playing *futbal*.

"Mincho, we are so far behind schedule already. Are you crazy?" I asked.

But after a day of sport and important team building, the volunteers showed up in force with renewed energy. They quickly completed the rest of the school and I had learned an important lesson – properly engaging the community is very important.

His work requires him to be away from his family for weeks at a time. During these times, he typically stays with different community members and quickly becomes part of the village's society. His travels have allowed him to learn from each community and the people are always hungry to hear him share the best practices he has seen in farming and livestock care. In essence, he has become a walking "Farmer's Almanac." I remember sitting around the fire late at night watching him instructing the community how to roast eggshells in the fire so that they can be fed back to the chickens. He explained to his eager listeners how the feed can replenish important nutrients and keep the chickens healthy as he showed them how to roast the shells. "After you feed your chickens the roasted shells, the new eggs won't break, even when you drop them," he bragged.

But his heart may be even larger than his personality. He loves helping his countrymen, working with them on development projects to remove the barriers to their dreams. As he moves around the Highlands, he always stops by a community that he has previously worked in to renew his relationships and follow up on the project's performance. Through this process, he has come to know which infrastructure and social systems work and which ones don't.

Over the years and many projects, Mincho and I have become more than friends. We are now brothers. Our skills complement each other's perfectly. I provide the technical designs while Mincho's construction lessons learned always lead to better, more sustainable solutions. We highly respect each other's role, knowing how important the other is for the overall success of the project. We feel no project team is complete without our "other brother."

We had already been working for several days to excavate the foundations of the new bridge crossing over the Rio Motagua. I knew that the rock foundation was going to be a challenge based upon what I had seen during the site visit. We had decided to do the excavation work early in a preliminary phase of the construction to confirm that an acceptable foundation solution could be found before completing the design. The work had been grueling with the team using ninety-pound jackhammers and rock chisels to do the excavation. The rain was not helping.

The rain was not the only surprise we encountered. The rock on the north side of the river proved to be a hard conglomerate that bent the rock drills and broke the steel tips on the jackhammers. Initially, the rock's weathered surface had given itself up to the pounding of the jackhammers, but then the brown rock had resisted, getting harder and harder as the work progressed. The excavation team had resorted to "double jack" mining methods when the jackhammers had no longer been effective. Double Jack mining is where one person holds the rock drill while two other members of the team swing twelve pound sledge hammers to drive the drill.

I watched as a sledge clanks as it hits the drill and the drill holder rotates it a quarter turn to allow the rock dust to escape just before the next sledge hits. Each blow results in the drill advancing a tiny fraction of an inch. This is repeated for hours at a time without a rest.

The work is slow but also intense – one missed swing by the sledgehammer and serious injury could result in the drill holder. Each man is riveted to his duty and not a word is uttered amongst, or to, them, so as not to break their concentration.

Despite the rain, progress was being made. I knew that the tremendous effort that was needed to excavate the rock also was the reason that the rock would be such a good foundation for the bridge and I was happy.

The rock on the south side of the river could not have been more different. The rock was volcanic, or igneous in nature and stoped, or laid in layers, steeply sloping towards the river. It was fine grained and a consistent gray color throughout its formation. The excavation team had cheered when the first large pieces of the rock had broken away with only a few minutes of the jackhammer's work. It was a welcome change to the slow progress on the other side.

Although the progress on the south side was good and spirits were high, I was concerned. The same reason that made the rock easier to remove was also the reason that it would not be sufficient to support the bridge – once loaded by a bridge, the rock could simply break away in chunks similar to those removed by the jackhammer, and the bridge would fall into the river.

The issue was serious enough to jeopardize the entire project. and I was stressed. If the rock was not suitable, the bridge would need to be moved to a new location – but there was not a suitable crossing location upstream or downstream for miles. I kept these worries to myself, knowing the fourteen communities that were hoping to use the bridge to access schools, health care and markets would be worried beyond imagination if I betrayed my uncertainty. They had hoped and prayed for a bridge for generations and the thought of crushing their dream was something that I shuddered to even think of.

How could they stop now after everything that had been done and the hopes were so high?

How would I ever be able to look into their eyes and explain that a bridge was not possible?

I had not slept well for days and the concern had completely distracted me from the mounting weather. My hopes rested on a white rock seam of granite flow located twenty feet below the volcanic rock. I had seen the granite seam when inspecting the face of the river bank. If the granite was wide and long enough to accept the bridge foundation, then it just might be the solution we needed.

The rain was never ending and now seemed like part of the expected environment. The river had risen over ten feet and had washed away many of the rocks gathered by the community members to assist with the construction – wasting the hundreds of hours spent by the community to gather them. The community knew that the effort would need to be repeated once the rains stopped. But they were not dismayed. They considered it only a minor setback to their goal of a new crossing across the dangerous river they called "The Assassin."

Now the rain began to come in sheets and demanded my

attention. The wind also began to howl, making a groaning sound as it passed through the pine tree tops. The workers tied themselves off to the bank in case a slip might put them in the river and wash them helplessly downstream. A new danger had arisen as full trees were now washing down the river having lost their home on the banks as the river scoured the soil away around their roots. One tree approached the worksite and nearly swept one of the workers downstream with its branches.

As the dangers mounted, Don Mike and I began to wonder if the work should continue. The equipment had been rented for a month and it might be months before it could be scheduled again. The rest of the bridge's construction had been planned for January, the driest month of the year. It had been selected to minimize the risk of the river washing away the falsework needed to hold up the forms while the concrete would harden. A delay of even a month would require the project to be delayed a full year till the next dry season. We stopped work and stepped back to do a careful safety evaluation of the site. But without electricity or a radio, we lacked the benefit of a current weather forcast. Based upon our experience, we determined that the risks could be mitigated, so the work pressed on.

Juan was one of the workers using the jackhammers. He was a strong, stocky man in his early twenties and has Down's Syndrome. He dressed in a faded red shirt with holes torn in the sleeves and his well-worn blue jeans were held up by a rope belt. On his feet, he wore shoes with soles made from recycled motorcycle tires and a soggy, worn New York Yankees baseball cap perched on his head. His spirit was always upbeat and cheery, and it seemed he is one of those people who has never had an evil thought enter his mind.

Even after several days of hard work, I still marveled at his energy and cheerfulness. His spirit lifted the rest of the team, including me. He thrived in the jackhammer work and how he could do "a man's job" under the approving smile of his father. But Juan's enthusiasm was proving to be dangerous as he attacked the rock without concern for his or his coworker's safety. With the rising waters of the river, Don Mike knew that something needed to be done.

He pulled Juan's father aside and explained his concern and the man's face darkened knowing that his son would be devastated with the news of being removed from the jackhammer duty. Don Mike stopped the work and asked Juan to come out of the rock cut. At first, Juan protested, saying that the work was going great and he was not tired – no break was needed. He then realized that it was only him who was being removed from the excavation and his head bowed in disappointment.

As he reached the top of the cut, tears were welling up in his eyes as Juan approached Don Mike and his father. I felt terrible, knowing Juan had finally been accepted as a member of the Team – and now he would be told he could not participate like so many other times in his life. His coworkers looked on in sadness. They knew it was the right thing to do, but they so hated to see the young man's spirit crushed.

Don Mike put his hand on Juan's shoulder and asked him to raise his head and look at him. Don Mike's warm smile somehow raised Juan's spirits and he managed a weak smile back. Don Mike then thanked him for his hard work, saying that he has never seen such a good jackhammer worker in all his years – but he had a more important job for Juan if he would accept the assignment.

He explained the danger of the passing trees to the workers and asked if Juan would serve as a spotter for any trees coming down the river and yell out a warning to his teammates. Don Mike looked Juan in the eye and said, "You are the only worker I can trust with such an important assignment and I really hope that you will accept."

Juan's face brightened. Of course he would accept this important assignment! He perched himself on a rock outcrop to overlook the upstream portion of the river. Juan's eyes never left the river as he dared not even allow himself to wipe away the rain from his face. He diligently scoured the landscape for danger and yelled a warning to the workers when even a small tree approached. He was doing important work, and nothing would distract him. Such was the heart of Don Mike and I loved him for it.

After several days, the jackhammers approached the white granite layer with the workers cheering as each rock chunk the size of a hardhat fell away under the rat-a-tap-tap of the jackhammer. As the jackhammers hit the granite, I knew we had a chance as the sound of the hammers changed under the resistance of the hard rock. The

Workers jackhammering the foundation excavation for the bridge.

cheery nature of the workers changed as well as they grimaced and doubled their efforts to remove even the smallest chip of the granite – but I gave a sigh of relief. The bridge project would move forward now firmly resting on a solid foundation.

The team ended its work for the day and the visitors retreated to their makeshift shelter while the community members returned to their homes. The shelter was made of corrugated metal roofing, rough sawn logs and poles and was perched on top of the north bank of the river. It not only held us, but also the tools and materials needed to do the construction.

The smell of diesel fuel, wet clothing and freshly cut wood filled the air and made an interesting, but not unpleasant, atmosphere in the room. No electricity was available for miles and the makeshift shelter had no windows, so the darkness was disturbed only by the candlelight that flickered back and forth creating shadows on the walls and across the people's faces. It all had a calming effect on us and over the weeks of our work, the little shack had grown to be a temporary home.

But now, the rain and wind shook the shelter and demanded to be noticed as it interrupted our nightly routine. We feared that the shelter might be blown away as the walls moved and the roof groaned under the strain of the wind. The thunder increased in volume and now seemed to never end.

"I've never heard thunder so continuous before," I shouted to Don Mike.

He replied in a loud, but firm, untroubled voice, "That is not thunder, my friend. It is the boulders rolling down the river that make that noise when they tumble into each other."

I stared back at him in disbelief. "Is that really true?" I said.

Don Mike motioned for us to go outside and he pushed open

the flimsy door to the shelter to go out into the storm and look. Immediately, the wind and rain hit us and caused us to pause next to the shack, but we soon regathered ourselves and pressed forward. In the twilight, I then saw the boulders – some the size of a pickup truck – tumbling effortlessly down the river as if they were toys. The steady stream of boulders parading down the river created a continuous roar as they rubbed against each other.

The river had turned mean, ugly and violent as it powerfully raced down the valley. The water had risen another five feet in the last few hours and now appeared that it might even threaten our shack. We walked back to the shelter, pressing forward against the wind and rain.

For the first time, I was now scared. The project had distracted me from the elements, but they could not be ignored now. Don Mike quietly walked over to his bed and slipped additional concrete blocks under the boards that supported his bedroll without saying a word. His quiet demeanor had a calming influence on me and the group. During his life, he had been in some tough scrapes and this storm did not seem to faze him in the slightest.

Soon there was a knock on the flimsy shack door and Rolando entered. Despite a full day's work in the elements, he did not appear tired in the slightest. He was accompanied by his lovely, wide-eyed daughter of maybe six years of age. They were bringing dinner that consisted of the same items each night – corn tortillas, black beans and hot Guatemalan coffee. Tonight, they also brought a special treat – a fried egg for each of us! A typical person living in the Highlands might expect to enjoy an egg once a week, so excitement filled the candlelit room.

But Rolando's usual cheery disposition was replaced with a serious look of concern. A community member had returned from town and reported the news that the Highland region had been devastated by the storm – which we now learned was named Hurricane Stan. Since the area lacked electricity and phone coverage, news was always in high demand. It was the obligation of every traveler to stop along his or her way and give an account of the news to those along the journey. This time, the news was incomplete at best, but it was clear that the country was hit hard and the devastation was extensive.

Don Mike's forehead wrinkled, and he frowned. He had experienced disasters before, earthquakes, hurricanes and The Violence. He knew how vulnerable the Highlands were and he was concerned. He explained to me that when communities live by subsistence farming, it is unlikely that reserves are stored to mitigate such disaster events. He listened intently to the report and then announced that he would try to return to town the next morning to get a better report on the damage.

The next morning, we finished up our breakfast by candlelight. We would head back to town with a dozen community members to assist us in the journey, leaving Mincho and Adam to finish the rock excavation.

Don Mike's 1983 Toyota Landcruiser pushed through the mud, rocks and debris along the road. The white Landcruiser was a trademark of sorts for Don Mike. It had been through many a treacherous journey with him, where he had even rolled it over twice on separate occasions. Over the decades, the Highland people had come to recognize the white, battle-scarred Landcruiser that contained its crazy Brit driver.

The vehicle had never failed Don Mike over the years, and he was confident that it would not now. When the debris could not be traversed, the community members would cut a path through, or around, the blockage using machetes, hoes and picks. The progress was slow but steady.

Finally, we traveled the sixteen kilometers to the town of San Jose Poaquil and upon our arrival, Don Mike spotted a bus at the station. He quickly jumped out of the truck and ran for the bus thinking it was getting ready to depart and knowing it could easily be hours before the next bus came along. He shouted back that he would learn what he could and report back to us in the next day or two as he lumbered up the street with one hand keeping his British flat cap on his head.

It was getting late, so we quickly turned the Landcruiser around and started the return journey to the bridge site, hoping to complete the trip before nightfall. As the shadows lengthened, it became harder and harder for me to navigate the route. I crept around the road's sharp corners and looked over the edge at the near vertical drop-offs knowing one false move would lead to certain death. Finally, I made it around the last bend and then realized that I had lost all the feeling in my hands because I had gripped the steering wheel so hard.

Adam and Mincho were relieved to see me enter the shelter just as complete darkness fell. Adam now realized that he should get word to Katie as she is probably worried sick about him given the devastation of the storm. It also dawned on me that my wife has probably heard the news of Hurricane Stan even in Wisconsin and probably was concerned. But, there was nothing we could do in the shack as we were so isolated from the rest of the world. We both wished we had asked Don Mike to get word back to our spouses that we were okay and doing well.

The next morning, the flimsy shack door burst open just before daybreak. It was Don Mike. We stared at him in disbelief. He was soaked to the bone and covered in mud. He told us the city of San Jose Poaquil is isolated with all roads into and out of the city being cut off by landslides. The bus he had raced for was parked, unable to go anywhere. Don Mike had decided to return to the bridge site and had walked all night, covering the sixteen kilometers back to the shelter in total darkness and the rain. He was completely exhausted and settled down to enjoy his hot cup of Guatemalan coffee.

As he eased into his chair, he tells us that he had learned little about the extent of the damage while he was in town. Since the town was isolated, little news had reached it and the radio stations were all silent, probably due to their towers being blown over. What little news he was able to hear was not good. It sounded like the entire country had been impacted by the storm, that stalled over the Highlands region. Roadways were closed by bridge washouts and landslides. Even the Pan American Highway was closed.

As dawn broke, The Team inspected the rock excavation. We determined that both sides were now ready to accept the bridge foundations and the work was deemed complete. The Team could demobilize. Rolando informed Don Mike of an alternate route that he had explored on his horse. It would take them up a small stream and bypass the roadways that were blocked by the landslides.

So, we set off on a journey back to Antiqua. Once again, community members accompanied the Landcruiser to assist it through the debris and landslides. We traveled mainly in silence as we saw the devastation of the hurricane. Home after home was destroyed by flooding and landslides and nearly all the potable water lines were damaged. The people we passed on the road looked like zombies as they blankly stared out into space as the Landcruiser pushed on. They were in shock and they seemed oblivious to their dirty and torn clothes. Most had not even taken the time to wipe the mud from their faces.

As we arrived back at Adam's home in Chimaltenango, Katie raced out of the house and nearly squeezed Adam in half with her hug. She was so worried about him and the Peace Corps kept calling every hour asking if she had any news of his condition – which made her even more afraid. After giving the Peace Corps office the good news on Adam's arrival, she handed the phone to me and I called Cathy, my wife. Of course, she was concerned, but said, "I knew you would be okay as long as you were with Don Mike. If anyone could ride out a hurricane safely, it would be him."

As Don Mike and I traveled back to Antigua, we passed continued death and destruction. It was reported that over 1,600 deaths occurred during Hurricane Stan and many remain missing to this day.

I asked if I should stay to help, but Don Mike encouraged me to return to my home. "Now is the time for rescue. Go back home and get ready to return when we can put your engineering skills to better use."

So, I reluctantly returned to my home in Wisconsin while always thinking of my friends in Guatemala and their suffering. As I told the stories of the destruction I had seen, my wife, Cathy, knew that I had to return and help. "I can't help them with water and bridges, but I can lend them you," she said, giving me a big hug.

The next morning, I returned to my work at the CH2M office in Milwaukee. I stood in my boss's office explaining the situation.

"So you think you will need to return to help out?" he asked.

I nodded.

"Do you know when you will need to leave?" he asked with a wrinkled brow.

I had to admit that I did not know.

"Do you know how long you will need to be gone?" he pressed.

I admitted I didn't know, realizing how ridiculous my request was.

"Okay. Get to work on transitioning your work duties to your coworkers during this leave of absence. Let me know if you need any help, but I suspect they will all rise to the occasion and pitch in. Just keep us posted and please stay safe."

I then knew how lucky I was to have such an understanding boss who also had a heart to help.

That evening, the phone rang. It was Don Mike.

"The Highlands people need you."

"When?" I asked.

"Tomorrow would be fine," Don Mike said flatly without emotion.

Don Mike inspecting a damaged bridge over the Rio Motagua.

CHAPTER SEVEN
Hurricane Stan Disaster Response

"Disaster work is different than development work. Development work is all about the sustainability. Disaster work is all about speed and the race to save lives."
Michael Shawcross

The day after talking with Don Mike I caught the first available flight and returned to Guatemala to help him with the Hurricane Stan response. He waved at me to get my attention at the airport as I exited the building. The normally crowded greeting area had been transformed into a noisy, chaotic sea of people and confusion. Planes and cargo jets full of relief supplies choked the airfield as it was stretched to its limits. Thousands of workers were moving like ants to unload the supplies onto countless trucks that clogged the access roads. What was typically a well-organized system now looked like total chaos.

Over the years, Don Mike had mentored me on how to do community development work. As we worked on schools, bridges and potable water systems together, Don Mike taught me the importance of a community-driven project and listening to the needs of the people to provide a sustainable solution. Development work requires patience to ensure that the right project is being done for the right people at the right time.

Time and again, Don Mike would say, "Never do a project until the community is fully ready. We need to be patient and let things develop at the community's pace." I had learned that sometimes years are required for a community to work through the important planning and consensus-building phase. Pushing a project through that is not ready only weakens the community's ability to sustain it in the long run.

This was my first disaster response and, as we bumped along in his trusty Landcruiser, Don Mike explained that a different approach is required. He had experienced many disasters including hurricanes, earthquakes and war. He explained that one needs to "put on a different hat" - thinking and reacting differently. Response time is now the critical element as lifesaving food, shelter and water need to be provided as quickly as possible.

As I looked at him across the front seat, Don Mike already looked tired and stressed. He clearly had been working around the clock and had only slept for a few hours when he couldn't force his eyes back open. His clothes, hair and beard were unkept. A fact that he apparently did not notice, and if he did, didn't care about. He was focused and knew that the people of the Highlands were hurting.

He explained to me that the infrastructure solutions for disasters fall into three categories. Temporary repairs that provide lifesaving service as soon as possible. A final repair or replacement to address the problem long term and sometimes, a transitional response to fill the gap between the temporary and final solution. This is especially true if the final solution may require a long period of time to be completed.

The key is to know when to use which type of solution based upon the risks and the limited resources available. He explained that he needed me to identify the possible solutions in each category, provide an estimate of the cost and construction time, and help him make recommendations to the municipality as quickly as possible. Don Mike explained that these are the important decisions that disaster responders need to make every day.

Restoring Access for Relief Supplies

The local municipal governments were overwhelmed by the disaster as many of their office buildings had been damaged and their staff and their families suffered from the devastation. Their top priority was for help with bridges to restore the vital links to the communities so relief supplies could reach those in need as quickly as possible. As we met with the municipality leadership, they looked shell shocked.

The mayor said, "Our machines can push the mud and trees off the roadways, but it is the bridges that are the problem. They have to be inspected before we can allow any relief trucks to pass over them."

He was clearly exhausted as he continued. "If damage has occurred, I need you to design and build the shoring and repairs needed on the spot and reinforce the bridges when the communities are trapped with no access. If the bridge has been destroyed, a temporary bridge or route must be built until a final solution can be done. My people are hurting, and they need help."

He then looked Don Mike in the eye and said, "We have known each other a long time. I trust you and know you will do all you can."

The municipalities had asked for help from the federal government, but all available engineers were working around the clock on the major transportation routes in the country, especially the Pan American Highway. There simply was not enough engineering capacity to work on the routes from the highways to many of the communities. They needed creative and fast solutions.

Our first assignment was to restore access to a Highlands community of 25,000 located near the top of the mountain ridge. Reports indicated that the hurricane had caused many landslides in the area and many people had lost their homes and loved ones. Food supplies were damaged, and a helicopter was being used to airlift food and medical supplies, but helicopters can only move a limited amount of supplies with each flight. A route to allow trucks to pass was needed to move the tons of life sustaining materials before things got worse.

Don Mike and I worked with Mincho, Adam and our friend George, the bulldozer operator, who had been reassigned from the La Garrucha roadway work to assist with the disaster response. George also had a new tool, a small rubber-tired backhoe that could help remove debris in waterways where a bulldozer could not access.

As the Landcruiser drove to where the road headed up the mountain, we passed home after home that had been destroyed by flooding and landslides. Temporary shelters were being assembled by the families out of the debris and blue plastic tarps had been erected to provide some form of protection from the elements. Tension filled the air and the normally beautiful Guatemala landscape was scarred, dark and muddy. Even the birds seemed to have stopped their singing and it seemed eerily quiet. The air was filled with the smell of mud and decomposing flesh.

The people seemed emotionless as they worked in silence to do what repairs they could. Entire families, including children and grandparents, toiled in silence to restore some form of civility to their environment. Friends and family helped each other as they pulled together makeshift shelters and consoled each other. I could see their sadness and their hearts were heavy, but they had few options but to pull themselves up by their own bootstraps and make the best of the situation.

As we drove, I saw a grandmother struggling to lift a heavy log to be used as a temporary support for her makeshift shelter. I slowed down to help her, and Don Mike said, "Sorry, we need to keep going."

I looked back at him, surprised. How could he just ignore her and her struggles? This was not the Don Mike I had grown to love and respect so much.

He must have sensed my thoughts and said, "I know you want to help everyone right now, but we need to prioritize. If we stop for everyone, we will never reach the bridges and thousands will suffer for it."

I realized he was right. We needed to focus our efforts where they would help the most.

George had already begun clearing the landslides and debris from the roadway but had been stopped dead in his tracks by the first bridge. The accumulated debris plugged the opening beneath the bridge, forcing the water to find a new pathway, washing out the road.

"So, what do you think? Do you think we can salvage her?" Don Mike asked me as we surveyed the damage.

I took in a deep breath and dove into the mud, water and debris to get a better look at the structure. I emerged covered in mud after completing the inspection and determining that, although the bridge had been roughed up a bit, it could be put back into service if it could be unplugged. I was happy that I could use my engineering skills and knowledge to help and was thankful that the bridge could be salvaged. I had never known a time when my engineering skills were more needed.

The backhoe went to work along with a team of fifty local volunteers who set to cutting and removing the fallen trees. Once the bridge was unplugged and the river flowed back into its original bed, the roadway could be restored, and trucks could pass. The volunteers hacked at the trees with their machetes and moved the rocks with prybars and the backhoe's assistance. The sound of the chop, chop, chop of the machetes filled the valley as they cut through the logs. Hoes were used to pull the mud and rocks out from under the bridge so the backhoe could move the piles of debris off to the side.

With so many workers, progress was made quickly, and the bridge was unplugged in a matter of hours. The sun had even come out to help dry the mud and as the backhoe worked to coax the river back under the bridge, I felt an overwhelming feeling of satisfaction.

A feeling that was interrupted by Don Mike saying, "Okay, that is the first one. The next bridge site is a real mess."

Don Mike and Adam walk across a temporary bridge being constructed from logs and boulders.

We trudged up the mountain road to the second stream crossing. The bridge had been completely removed by the rushing water and debris. Trees, boulders and mud were everywhere and in a tangled mess that seemed impossible to sort out. There was nothing to salvage here and I began siting a location for a new bridge that could be built of reinforced concrete. George arrived and reviewed my sketches.

"We don't have time for that," he exclaimed shaking his head. "Let's build something temporary using what the hurricane gave us. I can move those boulders over to make a temporary abutment and we can use some of the trees to form the bridge deck. I need you to tell me how many and what size logs we need to support the trucks. Let's get going as we have a few hours of daylight left."

I then "got it." I didn't need to design a bridge that would last fifty years. I needed to find a solution that could be built today that would work safely. The concrete bridge would need to come later. This was the time when temporary solutions were needed.

Slowly, we wound our way up the mountain roads repairing and replacing bridges with the assistance of the local community members. I was in awe as we watched scores of people work as volunteers to do the repairs knowing that many had also lost their own homes and even family members. I doubted that I would be able to muster the same spirit at such a time.

Don Mike was in his element – organizing the people, gathering materials and tools for repairs and providing encouragement to the communities at exactly the time when they needed it. He never seemed to tire and his calm demeanor and compassion were exactly what the people needed.

The hard work was incredibly rewarding for me too. Never had I been able to use my engineering skills for such good. But, at the same time, I had never seen such death and destruction. The pain in the people's faces told the stories of hardship and despair which drove me to do all I could with every fiber of my body. As the relief trucks entered the village for the first time, the typically reserved Mayan people cheered, clapped and cried – all at the same time.

Under Don Mike's leadership, this picture was repeated time and again as the Team worked its way up the mountain roads to village after village.

Agua es la Vida – Water is Life

Once critical access to the village was restored it was clear that potable water was the next priority. The waterlines that served the village were fed by springs located miles away. The enormity of the situation was easily seen as we stood on top of the ridge and could see more than one hundred landslides that scared the once beautiful slopes. The landslides caused by Hurricane Stan had broken the water lines in many places and the people were drinking contaminated water until repairs could be made.

The people were forced to gather water from the river or mud puddles to satisfy their thirst. What made matters worse, the surface waters were heavily contaminated with dead animals and debris from the hurricane. The water was putrid and dirty, but there were limited options. The pueblo had no electricity and dry firewood was in short supply, so the ability to boil the water was limited. The race against cholera, a highly infectious disease that can lead to dehydration and even death, was on and Don Mike was determined to not lose.

We set out to walk the water pipeline from the water source to the village. The going was slowed by the fallen trees, mud and landslides. We slowly trudged up the mountainside, sliding back a step for every two taken and our feet grew in size due to the thick mud that stuck to them. Soon, we found where several landslides had removed the piping and interrupted the potable water service. The good news was the spring that fed the water line was still flowing strong with clean water, it just needed to reach the people. The pipeline repairs were designed and mapped, but it would take weeks for new trenches to be dug into the side of the mountain and the pipe laid. Weeks the community did not have.

Back in the pueblo, a hastily organized meeting was held with its leaders. The meeting was held at the community center and the group sat in a circle on folding chairs. Everyone was tired and leaning forward with their hands on their knees. They were wet, tired, hungry and covered in mud due to the nonstop work required. Clearly, their hearts were heavy with sadness and the enormous work that lay ahead of them.

Don Mike started the meeting by thanking them all for their hard work and dedication. The water line breaks and repairs were explained along with a need for a temporary solution to the community's water problem. They discussed other potential water supply options. Another source of clean water needed to be found that could be somehow temporarily supplied to the pueblo. One of the leaders offered up a spring along a creek located a few kilometers upstream of the pueblo that was on his land. "Could it be of use?" he asked.

The entire group headed for the river to determine if lifesaving water could be provided by the spring. We slowly made our way up the valley hopping rocks, jumping over logs and wading the stream as we progressed the few kilometers to the spring site. The spring was found bubbling out of the side of the mountain just above the river. It had been exposed by a recent landslide and appeared to be the answer to the community's prayers. After the spring was inspected, it was determined that the water was clean, and the flow was strong.

Then Mincho had a suggestion. If the water could be piped down the valley to the roadway below the pueblo, pickup trucks with barrels could be used to shuttle the water up to the community. The barrels could be emptied into the pueblo's water fountains, providing clean water to everyone.

I frowned and shook my head. I told them I liked the idea but knew it would take weeks or more to dig in a trench for the waterline through the rocky terrain.

Don Mike then suggested, "Who says we need to bury the pipe? Let's simply lay the pipe on top of the ground next to the river.

If a rainstorm washes out the line, we will simply pick up the pipe downstream and reinstall it. Remember, we are at the end of the rainy season and the risk of additional heavy storms is minimal."

I shook my head, disgusted with myself. Why hadn't I seen the solution that was so simple? I clearly needed to change my thinking and focus on solutions that would provide immediate help.

*A pickup with tank transports the spring water
to the community's fountains.*

A customer account for water pipe and fittings had been placed with the hardware store, thanks in part to a financial donation from the Denver Rotary Clubs. That evening, I determined the quantities of pipe and their size and called in the list to the hardware store.

The next morning the piping was loaded onto trucks and driven up the mountain roadway to the pueblo. As the materials arrived, the volunteers from the community quickly unloaded the trucks and carried the pipes up the river to be installed.

Tree trunks were cut and rocks rolled out of the way as they prepared as flat of a path as possible for the pipeline. By that evening, the people of the pueblo were dancing in the square as they were fetching water from the fountain pools. Positive things were finally happening.

This process was repeated again and again to provide water quickly to other communities. The water lines would be walked, and material lists developed. Each morning the materials list was called in and loaded onto a truck for delivery. The truck would drive as close to the site as possible, but most times the community members needed to carry the materials the last few miles to the repair site. They were happy to do the backbreaking work and you could hear the singing of the workers as the moved along the footpaths with the sacks of cement and heavy pipes on their backs. In most cases, clean water was flowing within two days.

"Two days might seem fast to us but try living for two days without safe water and you will realize that two days will seem like a lifetime," Don Mike had said. Seeing the need drove the team to continue our efforts around the clock, seven days a week. Sleep time was down time.

As the rewarding, but grueling work continued, Don Mike insisted that the team take two days off to rest and recover.

Mincho protested, saying that there were simply too many people that needed help and there would be plenty of time for rest when we were done.

During the disaster response, Mincho's creative ideas were a critical addition to the team as he worked tirelessly to help the Highlands people. Despite the sobering nature of the work, he somehow never lost his sunny disposition and positive attitude towards life. It seemed every time the work was about to break the team's spirit, he would break out singing another Spanish love song, which always brought a smile to our faces and a spring back into our step.

But Don Mike knew that the team was wearing down and a team that was sick or injured would be of no help to anyone – so he insisted on the two-day break. As we recovered, minor scrapes and injuries were now discovered that had been ignored and were properly attended to. We did our best to catch up on sleep and properly nourish our tired bodies. We had not realized how worn out we had all become.

Mincho returned home. The area where he lived had also been impacted by the storm and even his home had sustained some damage. Angelica, his wife, had encouraged him to go and help those who had been so severely impacted, knowing his skills were highly needed. He was concerned but knew that his family would be able to put things back in order with the help of his neighbors. Now, he looked forward to seeing them all again and reviewing the progress of the repairs.

As the two-day break came to an end, the team was surprised to see Mincho returning, riding on top of a large truck. It was filled with sacks of corn and he greeted them with his usual smile and a wave. We then realize that Mincho and his entire family have spent the two day "break" walking to neighbor after neighbor asking for them to spare whatever corn they could to help their countrymen in need. He had passionately told the story of the suffering he had seen and the amazing resilience of the people as they did their best to put their lives back together.

His community had also been impacted by the storm, but they

now understood that other parts of the country were suffering terribly. As the donations of corn piled up, one of the neighbors even offered the use of a dump truck to transport the food back to the work site.

Don Mike shook his head in amazement and smiled. *It is just how Mincho rolls*, I thought as I watched him scooping out the corn into the hungry people's baskets and pails.

Don Mike enjoying a cup of coffee at Donna Luisa's in Antigua

CHAPTER EIGHT
How To Pay For a Bridge

"Every partner needs to contribute. The project is not a hand out, but a partnership where each partner contributes what it can. Through this process, community ownership is developed."
Michael Shawcross

The design for the Rio Motagua bridge was done and the list of materials was shared by Don Mike with all the communities in a large meeting in mid-November. The crowd stood silently with hopeful faces as he read the list. More and more, they had dared to believe the bridge would become a reality.

100 tons of stones

19,852 board feet of lumber for the formwork

133 cubic meters of gravel

93 cubic meters of sand

717 sacks of cement

15,532 linear feet of rebar

150 gallons of gasoline for the generator

1,400 pounds of nails

The list of dozens of additional items continued in excruciating detail. As each item was read, I could see people's hearts sank a little more like a boat being loaded with stones. How would they ever come up with the resources for such a bridge?

As Don Mike finished reading the list of quantities, he looked up at the silent group who were now looking at the ground, fidgeting with their hands. He told them that he knew that the community and municipality had depleted any reserves they might have had on the hurricane response and he understood that they had little to give to the project. But he asked them to look within and see the possibilities.

Rolando then cleared his throat and stood. "Yes, the list seems impossible," he said. "But thanks be to God, we all still have life. We are many people and we are proud and strong. If everyone contributes what they can, the task will be accomplished."

He compared it to everyone bringing a grain of sand to the river and soon a beach would be built. He then proceeded with the items one by one.

They could gather the stones from the river using work teams. "Should the communities work separately, or all together in one group?"

It was decided to work together on Saturdays when everyone would come and gather the stones. If the river did not have enough stones, they would roll rocks out of their fields to the roadway and use the municipality's truck to transport them to the bridge site. Everyone could help. The young, the old, men, women and children. They would all tote the rock they could and contribute to the piles.

Then the sand and the gravel were discussed. These materials that were needed for the concrete could also be gathered from the river. The hurricane had deposited large sand bars that were now exposed after the rains had ended. Don Mike explained that it would need to be washed to remove all the dirt and then sifted.

"A large task that would go quickly with many hands," said Rolando optimistically.

The group then discussed who might have pails and shovels to perform the task and several men discussed how they would build the large sieves needed.

Next was the wood. If each community contributed two pine trees, the wood order would be satisfied. Rolando would give each community a list that had the size and number of boards that they were to provide. The trees would be cut into boards where they fell on the mountain using a chain saw operated by one of the community members.

Everyone knew of his skill and felt blessed that he was able to do the work. Cutting the boards on the side of the mountain would allow the people to carry the wood to the roadway board by board instead of moving entire logs to a sawmill. Again, everyone could participate even if two or three children might be needed to move a board. A pickup could be used to move the wood to the bridge site.

A community logger cutting the boards using a chainsaw.

The items that could not be gathered by the community would need to be purchased by the municipalities and Engineers Without Borders. This list was divided up by the three parties with each party now knowing its responsibilities.

In the end, everyone had given what they had and a little more. It would not be easy for anyone, but the impossible now seemed to be possible. The meeting broke up with hope back in everyone's hearts. The silence was gone, and the room filled with chatter about workdays and details of the effort at hand.

After the meeting, I looked at the list of items to be provided by Engineers Without Borders USA. I knew that everyone had done their fair share and more, but the list still seemed overwhelming. The funds that would be needed were more than double that which had been raised for past projects and it was the most expensive project that Engineers Without Borders USA had done to date. We needed $60,000.

During the spring, the students had gone to the donors of past projects and sought commitments. Southminster Presbyterian of Waukesha, Wisconsin, had always been a generous donor of past projects. They agreed to do the same for the bridge and do a special collection to increase the contribution. All the other donors agreed to do what they had done in the past and maybe a bit more. But the funding was still $25,000 short.

The student design team had been writing grant applications all summer. Countless hours were spent in the library doing research on potential donors. Letters were written and a small prayer was said as each envelope was licked and placed in the mail. Slowly, the responses came back.

A typical one: "We were impressed by your application and the impact of your project, but we are unable to support your bridge at this time. We wish you the best of luck on this impressive project." Despite their tremendous efforts, no additional funds were found.

"Well, that was sure a lesson learned." James said, shaking his head.

Yash is a friend of mine and an engineer. One day over coffee, while I told him about the project, he suggested that Rotary clubs might be able to help. He was, and still is, an active Rotarian and committed to their motto of "service over self". He organized for the EWB Team to give a presentation at his club – Milwaukee's North Shore Rotary Club. None of us knew much about Rotary, but all avenues needed to be explored.

The club meeting was held in a restaurant north of Milwaukee. The students and I nervously picked at our lunch plates waiting for our time to give the presentation. When the time came, I made my way to the front of the room and kicked off the presentation by outlining the project and its details.

As I looked out over the audience, I could see the eyes of the crowd glaze over as they finished their dessert. Well, it was worth a try, I thought as I sat down and turned the rest of the presentation over to the students.

As the students passionately told the story of their visit to the community and the need that they saw, the crowd could not help but become engaged. Soon questions started to fly and the student's heartfelt answers continued to build the energy. The moderator had to end the discussion due to time but as the meeting flowed out into the hallway the Rotarians continued to ask questions and the students continued to tell their stories.

Soon a check from the club arrived in the mail. It was for $500 plus another $142 from the free will collection made at the door. We were on our way and I realized that my best contribution would be to "get out of the way" and let the students tell the story with their energy and passion.

Yash introduced us to other clubs and soon the bridge team was giving presentations to clubs all around Milwaukee. This included The Rotary Club of Milwaukee, the largest club in Rotary's southeast Wisconsin District. Not only was support secured for the bridge, but the groundwork was laid for a long-term relationship between the club and all the Wisconsin student chapters of EWB-USA that continues to this day.

Over the years, students from the university chapters would come to the club meetings and present their projects. They also learned about Rotary, its mission and vision, which are in complete alignment with Engineers Without Borders USA.

The two volunteer-based groups were a perfect match. Both had a culture of "service over self" and a global view. The Rotarians brought experience and mentoring to the students while the EWB students brought a youthful passion and energy to the clubs. Over the years, many volunteers have become members of both organizations, including myself.

The students also needed to raise the money to pay for their airfare. The Marquette Chapter of EWB had used "the thousand letter campaign" successfully in the past where each member of the chapter would send a letter to friends and family asking for a contribution of $50, many times in lieu of a Christmas present. They also sold grilled cheese sandwiches to their fellow students during final exam week. Somehow, they found the time to do their own studying and make the cheese sandwiches that always seemed to be a big hit with the sleepy studiers.

Moving Money

The donations were carefully tracked penny by penny as we inched towards our goal. James reported, "We are getting close, guys! Let's keep up the letter writing and keep making those sandwiches!"

Finally, the goal was met for the materials. I went to the bank to wire the funds to Don Mike's account. We struggled through the Spanish translation for addresses and came to the question on the form – "purpose of fund transfer?". I responded, "to build a vehicle bridge." The banker looked over her glasses at me and then realized I was not joking.

"You don't hear that every day," she said with a smile.

As we completed the paperwork, I told her the story of the project and its impact on the people. As we wrapped up the transaction, she smiled and said, "I think we can waive the fee for this one. Good luck and send us a picture."

Don Mike patiently waited for the news that the funds had cleared the banking system verifications. Finally, he received the word and was ready to make the purchases. But how would the funds be transferred to the hardware stores? In later years, electronic transfers would be the norm, but not in 2005. It was strictly a "'cash on the barrel" business.

What made matters worse was that no banks existed near the project site, so cash would need to be taken to the site to make the purchases. The largest note available at the time in Guatemala was 100 quetzals – or about $15. When Don Mike went to his bank in Antigua, the teller counted out each stack, consisting of 100 bills, and piled them up on the counter. The stacks were checked and double checked by the bank tellers and then checked one more time by Don Mike. Altogether, there were 4000 – crisp 100 quetzal bills neatly stacked on the counter.

Finally, after several hours, the lead bank teller peered over the stack of bills and asked Don Mike, "Will there be anything else?"

Don Mike smiled back and thanked her as he stuffed the piles of money into an old blue backpack.

The blue backpack had been selected because it was old, torn and non-descript. Surely it would not attract any attention of even the most observant robber. Don Mike lugged the backpack out to the Landcruiser and gently nestled it between us in the front seat.

"I never knew the money for a bridge would be so heavy," he said with a smile as he rested his arm on the backpack giving it a gentle pat. "I just hope the seams hold out on that bag. Can you imagine what would happen if they burst and spilled the bills all over the street?" We both laughed at the thought of the commotion it would cause.

We stopped in San Martin at the last gas station to fill the thirsty Landcruiser and stretch our legs before the heading onto the remote portion of the journey. Don Mike nervously looked around to make sure that we were not being followed by a bandit who may have gotten word that we were toting the cash for a bridge with us.

We went into the gas station to pay the bill with Don Mike carrying the backpack, trying to make it look as light as a feather instead of the heavy weight it was. We then returned to the truck and headed on our way down the bumpy, dusty road to the bridge site. Our spirits were high and, as was the custom, we sang songs to pass the time since the radio had ceased to work long ago.

Then, about an hour into the journey, Don Mike was looking for his water bottle in the back seat when his face darkened. "I don't see the backpack." He said with a start. "I could have sworn I put it back here after we stopped for gas." He started to climb into the back seat, throwing items around the cab haphazardly looking for the backpack.

Finally, I stopped the Landcruiser and joined in the search as we frantically looked for the backpack. Then, we stepped back, and a more structured search occurred lining up along the highway every item removed from the Landcruiser. Still, the backpack failed to show and Don Mike looked like he could hardly breathe.

Both of us could not help but think of the consequences if the bag was lost. What would we tell the donors? What would we tell the communities? The communities had already done much of the work with the stones and wood piled near the site. The thought made us sick and we repacked the Landcruiser in silence lost in our desperate thoughts.

As we sat on the side of the roadway trying to gather our composure, we racked our brains on where the backpack might be. Our only conclusion - the only hope - was that it was somehow left at the gas station. We hastily turned the Landcruiser around and headed back towards the gas station. As we sped back the Landcruiser was hitting only the tops of the bumps as it skipped down the roadway at high speed. Still, the minutes were passing like hours and Don Mike kept pressing me to drive faster.

The vehicle had not come to a complete stop before Don Mike leaped out and raced towards the cashier's station. He looked around and saw no trace of the backpack He then gathered himself and tried to calmly ask the young lady behind the desk if she might have happened to see a blue backpack that he had left there three hours before. The bag easily carried thirty times the young lady's annual salary – a temptation that would likely be too great for even the most honest of people.

She smiled back at him and said, "Why yes. I figured you would come back for it." As she pulled it from behind the counter. "What do you have in this thing? It sure is heavy." She remarked. Don Mike snatched the bag out of her hands and thanked her without answering the question. He turned his back and opened the bag to find all the bills stacked orderly inside. He was nearly overcome with emotion.

"Is there anything else I can do for you?" she asked looking at Don Mike with a smile.

He did not know how to respond. He didn't think it would be wise to tell her the story, yet he wanted to thank her for her honesty. He hastily grabbed two bottles of Coca Cola, leaving a large tip and one bottle on the counter while he handed the other to me.

Back in the truck, we pulled around the corner and Don Mike hastily counted the money. It was all there. She must have placed the backpack behind the counter, resisting the temptation to look inside at what could possibly be so heavy.

Not a penny was missing.

From left to right, The Author, Don Mike and Mincho at the plaque of the completed Rio Motagua bridge. Photo credit Kevin Hagen

CHAPTER NINE
The Bridge Construction

"Safety is the priority and cannot be compromised.
Every worker deserves to go home in the same
condition they arrived at the construction site."
Michael Shawcross

Don Mike had searched high and low to identify a place for the team to stay. Each of the fourteen communities had offered space in a home or school, but the distance from the bridge site would make it difficult to move people and tools efficiently. He finally suggested that the work team camp on the bank of the river.

I contemplated the idea as we had never attempted it before. It certainly would complicate things since the site had no drinking water, electricity or shelter, but staying near the construction site certainly had its advantages. We could keep an eye on the materials and tools at the site and the commute would require only a few steps. Finally, I agreed and said. "Let's go for it. It can only add to the adventure."

If an army "marches on its stomach," so does a bridge crew. One bad meal could shut down the entire operation – a risk that must be avoided. As we contemplated the situation, Mincho's wife, Angelica came and approached Don Mike. Angelica is a short, stocky woman with flowing black hair, a laugh that immediately puts anyone at ease and a hug that could squeeze a normal man in half.

She explained that for years she had seen the projects Don Mike had worked on with her husband and felt that she could only contribute moral support with best wishes and prayers. She now saw her opportunity to contribute more with her skills and asked if she could cook for the work team.

Of course, he agreed, and she clapped her hands together with joy. Not only would she be able to contribute with her skills as an excellent cook, she would also be with her husband during the construction. The kids were old enough now that they could stay with Grandma, so everything was set.

She reviewed her list and described what would be needed to set up the camp before the construction could begin. Time was running short, as construction was scheduled to start in a month. Barrels were needed to store drinking water and of course, we would need to bathe in the river. Large tarps could be used to make a dining area and a makeshift kitchen. Men's and women's latrines needed to be dug and outfitted with a toilet and hand washing station.

She then described the mountain of firewood that would be needed to cook for the team. Don Mike suggested that maybe a propane gas stove might be easier and more efficient. Her eyes lit up.

"Would that really be possible? I have always wanted to cook over a gas stove," she said. The thought of not having to breathe in the wood smoke and worry about the risks of an open flame had only been a dream of hers up until now.

Angelica's Stove

We drove to the appliance store and Angelica was overwhelmed. She talked with the salesman and reviewed the selection of stoves. Each was fueled by propane gas and would sit upon a table, similar to a camping stove. There were dozens of stoves. Everything from one to five burners, gadgets and colors galore. They were so shiny and new that she dared not even touch them. Finally, after checking out each stove several times, she came to me and asked which one she should purchase.

I laughed and said, "Angelica, I know you would never tell me what concrete mixer I should use for the project and I would not even dream of telling you which stove we should buy for you. Please select the one you want and don't worry about the price."

Angelica frowned and went back to fussing over the sparkling new stoves and stressing over the decision she must make. She finally selected a four-burner white stove with a fold-up cover and warming station on its side. As she stood next to the stove, allowing herself to finally touch the glossy finish, she beamed with a smile that filled the room.

With the stove purchased, we drove across town to buy the food that would be needed, and Angelica was clearly in charge. We went to the market area and she bartered for large bags of beans, corn, carrots, onions and other vegetables. Each vegetable was carefully reviewed under her discerning eye and any item with the slightest flaw was returned to the pile despite the protests of the vendor. Large gunnysacks were filled, and the men struggled to haul them to the pickup trucks nearby. After double checking her list and making one last stop to the spice area of the market, she said she was ready, and we could go.

The final stop would be the livestock market. There would be no electricity at the site, so a refrigerator was out of the question. To keep any meat from spoiling, live animals would be purchased and brought to the site where they could be slaughtered the day that they would be cooked. Angelica carefully selected the chickens she would use, and Mincho negotiated for a nice sized pig.

It was quite the sight as we pulled out of town with several pickup trucks loaded to the brim with food, cooking utensils, chickens and the pig. I was riding in the back of one of the pickup trucks with the chickens and pig with the task of keeping an eye on them.

Soon I started to panic when I saw that the string that tied some of the chicken's legs together was coming loose. I feverishly tried to quickly retie the string, but the bumpy road made it impossible and soon the entire string of eight chickens were loose.

The chickens quickly made their escape and the truck screeched to a halt after my desperate call for help. The pig excitedly cheered on the chickens with loud grunts as everyone frantically chased the chickens on the mountainside. As we tied the last chicken to the string, we wiped the sweat from our brows and laughed. This was going to be an adventure!

Groundbreaking

Nearly a thousand people came for the groundbreaking ceremony. They came, walking down the mountain paths with children and grandparents in tow. It was a perfect, sunny day and the smiles were so broad that they could not help but fill the entire river valley with joyful energy.

The women of the communities had also been planning and worked together to prepare the celebration meal. Each was dressed in her finest blouse of traditional Mayan weaving, proudly representing their community. Although resources are scare, it is very important to the community that such an important event is properly celebrated. Every family contributes what they can in a "potluck style" celebration.

Pots the size of bathtubs were full of soup, stocked with vegetables brought from countless gardens. The cooks cut up the vegetables while others stirred the pots over the open fire with large spoons that resembled canoe paddles. The noisy chatter and laughter could be heard over the *pat, pat, pat* of the tortilla makers who skillfully roasted the tortillas over the open coals.

The men stood on the riverbank and pointed at the excavation for the footings and studied the large piles of stones, sand and gravel that had been gathered. They discussed the work that would be required and guessed at how long the project might take to be completed.

The students and other volunteers soon arrived with Don Mike and were engulfed by the crowd who wanted to catch a better glimpse of the foreigners. Initially, the volunteers stood in a tight circle while the community stared and pointed at the strangers from the north.

But soon Don Mike broke the ice by introducing the volunteers to the leaders of the communities and the volunteers were absorbed into the sea of people through handshakes and smiles.

The speeches began with each leader taking his turn at trying to outdo the previous speaker with energy and passion. The ground was covered with green pine needles in the Mayan tradition and the smell of pine trees mixed with that of the wood smoke and cooking soup to form a fragrance that could not help but make one hungry.

The Mayan Shaman lights the candles for the bridge blessing ceremony

After the speeches, a Mayan shaman conducted a traditional ceremony. He carefully placed candles in a circle, with each color candle carefully placed in the correct position. Pine needles were also used around the perimeter and flowers were added for a colorful accent. The candles were then lit, and the shaman prayed as they burned. Then it was the Catholic priest's turn, as he blessed the work and workers while burning incense and sprinkling holy water. As I took in all the events, the Mayan and Christian rituals seemed so different – but, at the same time, right and proper.

Lunch is prepared for the groundbreaking celebration

Next, the soup and fresh tortillas were served. Somehow the group of tortilla makers were able to keep up with the crowd's demand and everyone settled on the bank of the river to enjoy each other's company and a wonderful meal. The sun warmed them as they relaxed on the bank after stuffing themselves. There was less chatter now as

they digested their meal and a few even dozed off to the sound of the
river babbling along through the rocks below.

I couldn't help myself and approached Rolando to see if it would
be rude if we took advantage of the large workforce and started work
before everyone left. Rolando gave a surprised look back and said,
"Why not?"

He made the announcement that work would begin on the bridge
in twenty minutes and he needed everyone's help. Everyone looked
surprised as they assembled along the banks of the river near the bridge
site. Mincho and Rolando then organized the people into the work
teams. Soon the crowd was passing stones for the construction of the
pier that would be built in the center of the river. Rock lines were
established on both sides of the river and passed the stones previously
gathered and stacked on the banks. Each stone was picked up from its
resting place on the bank and passed from person to person until it
was carefully placed into its position in the center pier.

The work team of hundreds of people working together was the
largest I had ever seen. Everyone was excited to see the work begin
and it seemed to solidify the group's unity that the bridge was for the
benefit of everyone. The large crowd made quick work of the task and
everyone stood back and proudly realized what fourteen communities
working together as one team could do.

We all now knew we could do it.

After lunch, the community moves rocks to build the center pier

Setting up Camp

James led the volunteers in scouting out the locations for the team's tents on the soft soil of the riverbank. The sound of the river would lull them to sleep at night and Angelica was hard at work finalizing the dining and kitchen areas. One of the volunteers remarked, "Wow, we even have a petting zoo!"

Don Mike then informed the surprised volunteer that the animals would be the protein source for the team during the build. It was clear that most of the volunteers had never had to prepare a live animal for their meal and they looked a bit shocked and squeamish.

Angelica and a few ladies from the community who helped her cook, would rise at four a.m. to begin making breakfast and setting the coffee to boil. The attachment the volunteers felt towards the chickens soon diminished as the roosters crowed long before the volunteers were ready to rise.

"Any chance we can eat the roosters first?" one volunteer asked Angelica in a tired voice.

On the first day, two local entrepreneurs set up a small store to sell snacks, soda and beer. The volunteers made sure their efforts were not wasted as a steady stream of shoppers from the work team came to buy a snack after dinner. The storekeepers even started to take requests for different types of snacks and treats to keep their customers happy.

Organizing the Work Teams

The construction of the bridge would require many activities to be done simultaneously. First, the foundations and abutments would be poured for each end of the bridge. At the same time, the pier made of stones and sandbags would be built in the center of the river. This was needed to temporarily support the wooden trusses that would hold the concrete for the beams in place until it hardened. Once the beams were cast, the concrete deck would be poured to provide the final driving surface.

Rolando met with the representatives of each community every day to organize the work teams. Each day, a different community was responsible for providing the workers needed. One day, I went to Rolando and explained that fewer workers would be needed the following day due to the tasks required. Rolando frowned and then explained that it was important to keep the number of workers the same each day to keep the effort fair between communities. I then realized that it was up to me to plan and effectively use the workforce.

Since the workers changed each day, it was important to have an orientation meeting each morning after the workers arrived. Mincho explained the tasks of the day along with the safety measures that would be needed for each task. The meeting needed to be translated from Spanish into the two Mayan languages and questions translated back. Care was also taken to do detailed demonstrations so that the workers understood how to do the tasks safely.

At the end of the day, the tired workers made their way home up the mountain carrying their shovels, picks and hoes. If the workers were from the south side of the river, Don Mike would give as many workers as possible a lift to their homes. As many as eighteen workers would pile into and on top of, the Landcruiser for the trip, which would be repeated as many times as needed until the workers were safely home. The workers were excited about the progress made and talked excitedly with each other, wondering how much progress would be made before their next turn to work came.

Bees!

As I laid out the exact location of the bridge footings and forms, I swiped away at a flying insect that kept pestering me. Another and then another appeared and after catching one with my gloved hand I realized that it was a bee. Looking around, I then realized that a bee's nest was living in the old rock wall adjacent to the location of the new bridge. Knowing that the workers would be exposed to the bees for much of the construction time, I was determined to remove the nest.

My first attempt was a smudge pot to smoke them out. I cut holes in a metal pail and placed embers from a fire in the bottom before covering them with pine needles and wet wood chips. The pail billowed smoke and the bees came out to identify the enemy that was upon them. After spending several hours trying to convince the bees it was time to leave, I decided to try another tactic.

I waited until dark and soaked a rag in diesel fuel, wrapped it around a long pole and lit the rag on fire. I then stuck the fiery end up by the bees' nest with the hope that the bees would attack the fire and kill themselves. The bees quickly responded to the threat and swarmed around the fire and me, as I tried to get as far away from the nest as possible. After several hours, I retreated to my tent. Surely the bees would be gone in the morning.

As the new day dawned, I frowned while looking up at the bee's nest. The bees were working hard like nothing had happened, happily flying into and out of the masonry hole that formed the entrance to their home.

The whole time, the community had been watching this battle of me vs the bees with great interest. I had not asked for help and was bound and determined to win the battle. The community was more than happy to sit back and watch the foreigner entertain them with his feeble attempts. Rolando finally came to me while I was seated on a stone and puzzling out my next move. He asked if it would be okay if he gave it a try and I was relieved to step back and let him give it an attempt.

Rolando then brought over a ladder made from two tree saplings and bravely climbed up to the nest. I was terrified and insisted that Rolando come down before he was attacked by the swarm of bees. Rolando calmly ignored my pleas and upon reaching the nest entrance, bravely removed the stones so he could reach his hand inside the nest. The bees swarmed around him, but somehow, he was unharmed. He then pulled out handful after handful of honeycombs and passed it down to his friends below, who immediately set to eating the sweet treat.

As Rolando came down the ladder, he winked at me and informed me that these were native bees and did not have stingers! They would pose no risk to the workers and might even give them a tasty treat now and again during the construction.

I shook my head and laughed. I had learned a valuable lesson and would engage the community in the tasks going forward.

Wheelbarrows and "Feeding the Beast"

Because of the hurricane work, George had not been able to complete the construction of the roadway on the north side of the bridge. Heavy trucks would not be able to access that end of the bridge site for several months, so all the materials were delivered to the south side of the bridge. As lumber, sand, stone and cement piled up, space was soon at a premium. Since everything was moved by hand, a mistake on placing materials in the wrong location might result in a day's work for ten men to move it. Careful planning was necessary.

Fortunately, the existing pedestrian bridge was still intact, but pretty shaky to say the least. It consisted of railroad rail beams that spanned the opening over the river with wood planks for the walkway. The abutments were made of bricks and carefully placed in a partial arch to reduce the length of the crossing. The bridge would bounce as the wheelbarrows passed over the walkway, much to the delight of any children who happened to be sitting on the deck.

We quickly discerned that moving half the construction materials to the north side of the river would be the "critical path" that would drive the schedule. Ten wheelbarrows were dedicated to the task of moving the material along with their "drivers" and shovelers to load the wheelbarrows.

The wheelbarrow crew worked from first light to dusk day after day, moving the material across the existing foot bridge to the north side of the river. The wheels squeaked as they were pushed by the men making a strange tune as the ten men and their wheelbarrows relentlessly pursued their task. It was the sound that greeted Don Mike and me as we finished our morning coffee and the sound that marked the end of the day as the sun set.

True to his nature, Mincho would jump in and grab a wheelbarrow to give one of the workers a break when it seemed the team needed some encouragement. He would push the heavy load and break out into one of his love songs, causing Angelica to blush as she was working to prepare the next meal.

The wheelbarrow team continues its seemingly never-ending task of moving material across the foot bridge.

And so, slowly but surely, half of the stacks of material made their way across the river to be placed by the builders. Concrete was to be mixed using small gasoline powered mixers. Five-gallon buckets would be passed along a line of workers from the stockpile to the mixer. Three pails of stone, two pails of sand and a half bag of cement was the recipe that would be repeated over 2,000 times during the life of the project.

It seemed like everyone would find their place as the young, strong men would establish themselves by the mixer to throw the buckets of material into the mixer, or "the beast" as it became to be known. The grandfathers and young boys would shovel the material into the buckets at the stockpile, which would then be passed down the line of workers towards "the beast" and its endless appetite. The students jumped in with enthusiasm, which quickly waned under the constant, heavy work.

"I can't believe they can keep this up for hours on end," Molly exclaimed as she passed a shovel to a grandmother so she could take a break.

Given the amount of concrete that needed to be poured and the small batches that could be mixed, the sound of the mixer filled the air from sunrise to dusk. Several times the last loads of concrete would be poured after dark as flashlights and lanterns illuminated the work. Slowly and surely, the wood forms were filled with the rebar and concrete as planned and the bridge began to rise up from the banks like a corn stalk in a field.

The final concrete for a foundation pour is placed
via flashlight and lantern.

The Hagen Truss

The most challenging portion of the work was the fabrication and placement of the twenty-two wooden trusses needed to support the concrete beams and deck until they hardened. Kevin, the truss designer, was onsite and assisted by Jim, a master carpenter who had led the carpentry teams on several previous bridges. James happily joined this team of grizzled veterans along with Marquette Engineering Alumni and carpenter Rob. Together, with the community, they formed a carpentry team that rose to the challenge of the task.

Most of the community members had never operated the skill saws that would be powered by the gasoline generator. Kevin and Jim relished the role of training the community volunteers in the safe use of the tools. Initially, the community members were afraid to use the power tools and insisted that the foreigners operate them. But soon they accepted the invitation extended by Jim and Kevin to learn how to safely use the saws. They soon gained confidence in their abilities under the ever-watchful eye of Kevin and Jim.

The first step was to carefully cut the 3,168 different wood pieces that would be needed to assemble the trusses like a puzzle. Each wood piece needed to be carefully cut and notched to fit into its specified place. The variable nature of the wood caused by the chainsaw operation used to cut the boards posed a challenge. Each piece was carefully "fitted" using a team of seven men with machetes who ensured that each piece fit snuggly with its mate. Molly and Andy carefully checked the size of each wood piece, sending any piece back to the machete team that didn't quite pass their inspection. The *whack, whack, whack* of the machetes filled the air as the *whirr* of the generator and skill saws provided the constant background noise.

Once the pieces were cut and fitted, they were turned over to the nailing crew. Each piece had a specified nailing pattern that would be needed to make the connection to its mate. Kevin provided a template for each nailing pattern and the nailing crew quickly learned how to make the connections. Now the hammers added their ringing sound to the carpentry area as the carpenters pounded hundreds of pounds of nails into the pine wood.

One of the more challenging aspects of the assembly was that each truss needed to be built with a slight curve, or camber to its shape. The camber was designed so that it would be reduced to nearly a straight line as the weight of the concrete applied to the trusses deflected it downward. It all looked good on paper, but I was concerned how consistently the trusses would settle under the weight given the variable nature of the wood, nailing and fitting of the pieces.

"Only one way to really tell," said Kevin. "We need to do a load test."

Two of the trusses were set up on the riverbank to mimic the installation over the water. The space, or "beam pocket," was formed between the two trusses and boards were placed between the trusses to form the bottom. Bags of cement were then placed to simulate the load that the beams would be under when the concrete would be poured. One by one, the workers added the bags to the beam pocket between the two trusses and the amount it sagged under the cement's weight was measured by the engineers. Soon over 300 sacks of cement were stacked between the trusses to simulate the 30,000 pounds of concrete that the trusses would need to support. The trusses groaned and settled under the weight, flattening out to the exact location as predicted by Kevin in his design.

"I don't think a fabrication shop in the United States could have done a better job," Kevin exclaimed proudly. Satisfied, the carpentry crew was ready for their trusses to be placed over the river.

Placing the Trusses

The plan for setting the trusses in place was to utilize a backhoe from the municipality to lift them up from their location on the bank and then swing them slowly out over the river so one end would rest gently on the temporary pier. This required the backhoe to work its way out onto the abutment of the existing bridge and position itself such that it could swing the truss into the correct position.

The backhoe arrived and began to get into position. It maneuvered back and forth on the narrow abutment being careful to not get too close to the edge. After two hours, the machine was finally in place and ready to lift the first truss into position. The bucket reached over to the truss as it rested on the bank and gently picked it up using a chain and a series of straps. I then motioned for the backhoe operator to swing the truss over into position. But the operator looked confused and refused to swing the truss. A series of hand gestures ensued, but clearly, we were not communicating.

Frustrated, I stormed over to the backhoe and climbed up into the cab to find out what the issue was. I learned that the backhoe's mechanism used to swing the bucket was broken. Placing the trusses using the backhoe was simply not an option. Another way needed to be found.

The volunteer engineers met with Don Mike and Rolando to brainstorm a solution. Ideas were suggested and systematically discarded one by one. Soon, the group ran out of ideas and disbanded to meet again. I couldn't sleep a wink, racking my brain for a solution.

When we met again the next morning, someone suggested we use the existing pedestrian bridge to support the trusses so they could be passed down from above to their position. The idea seemed risky as the existing bridge was supported by old, cast iron railroad rails that were used as beams. What made matters worse, the rails were spliced to reach across the span. The splices were typical railroad rail splices that consist of metal plates that are bolted to each railroad rail to make a connection. These splices are used to keep rails in alignment, not to transfer heavy loads like a beam.

Rob, Kevin and I took up the challenge and worked late into the night calculating the load that the pedestrian bridge could support, only to determine in the wee hours of the morning that it simply could not be calculated with any certainty. There were too many unknowns to really determine what load the bridge could safely hold.

I was now stressed beyond imagination. If a solution could not be found, the entire project was in jeopardy. For a second night in a row, I didn't sleep a wink and paced up and down the riverbank staring at the bridge site as if a solution could magically appear if I just willed it to happen. That night, as I crawled into my tent and tried to sleep. I promised God that I would never try such a challenging project again if He would help me out, just this once with a solution.

The next morning, Don Mike and Rolando came to me. I was so upset that I had failed to even get a cup of Angelica's amazing Guatemala coffee. They calmly asked me to sit down and relax as they poured me a cup of the freshly brewed black gold. Then Rolando explained his idea. It consisted of placing a column made out of a tree trunk between the pedestrian bridge railroad rails and the top of the temporary pier. Both ends could be notched for a firm connection and he had identified a tree of sufficient length to get the job done.

I stared back at them. We had considered a similar solution but discarded it since we lacked any timbers long enough to make the connection. But using the tree trunk was not considered and just might work.

I began to talk through the idea out loud, "Yes, I could check the size of tree trunk needed to hold the load. I would also need to check the lateral load that would be transferred to the temporary pier, but we could always add some bracing to support it." A broad grin came over my face and my heart lightened. "This might just work," I said slapping Rolando on the back.

I realized once again, these are the type of problems that Rolando solves time and again using his innovation and creative thinking. Simply calling for a larger piece of equipment like I would do in the

U.S. was not an option for him and he was used to thinking "outside the box" to find solutions.

With the calculations done, the tree was brought to the site and lowered into position using ropes. There were still plenty of unknowns in the analysis and one of the volunteers suggested that maybe the first truss should be set using community members only.

"I don't think we can risk one of the U.S. volunteers getting hurt," he said.

Don Mike turned red as he tried to control his anger. The safety of everyone was most important to him. It mattered not what country the person was from, and in fact, an injury to a community member who relied on their physical abilities to make a living to support them and their family was likely worse than an injury to an American volunteer. Don Mike slowly, and firmly explained this to the volunteer. He was ashamed because he knew Don Mike was right and sheepishly apologized, bowed his head in embarrassment.

The only logical approach was to perform a load test on the bridge. Material would be placed out on the bridge one wheelbarrow load at a time, starting in the middle and then extending towards the abutment. Only one person would be allowed on the bridge to place the load and I insisted that it be me. I put on my safety harness and checked the connection of the safety lanyard to a rope tied back to a tree on the bank. As I wheeled sacks of cement out onto the bridge, the railroad rails groaned and deflected, shifting the load to the tree trunk in the middle as planned. I tried to always keep a clear path back to safety in case the bridge started to give way and I needed to race back - but the bridge held.

Now we were ready for the trusses to be placed. The first truss was carried out onto the pedestrian bridge with Don Mike leading the way. The crew gently lowered the truss over the side of the bridge using ropes and slid it down the tree trunk to a crew on the pier who received it. The pier crew then slid the truss across the top of the pier, into the correct position. Everyone breathed a sigh of relief and smiles filled the faces of the team as the first truss was moved into position. Then the other twenty-one trusses were carried out and set in the same manner. For the first time, the new bridge could be visualized.

Workers placing the trusses that will support the concrete for the bridge.

Completing the Bridge

Now with the trusses in place the excitement began to build. For more than a week, rebar cages for the beams had been pre-tied on the banks of the river, ready to be set into the beam pockets created by the trusses. The next morning, eighty workers lifted the cages and slowly moved them onto the bridge like a centipede with 160 legs. Working in unison, they slowly lowered the steel into place and then repeated the process until all five beams were ready for concrete. With the rebar now in its proper position, the beams were ready for concrete.

The next morning, "the beast" was fed its meal of sand, stone and cement until it filled the beam pockets with concrete, using the wheelbarrows to move the mix between the beast and the beam pocket. It took a full eight hours for the beams to be poured and when the final wheelbarrow of concrete was placed as the sun set, everyone cheered.

The rebar for the deck was then tied in place in front of an ever-growing crowd who gathered each day to watch the bridge progress. Finally, a few days later, the day came for the concrete deck to be poured. With so many volunteers now showing up to the project, it was determined that three concrete mixers would be used simultaneously to perform the task.

The three mixers roared to life as their engines were started, and the bucket brigade lines passed the pails of sand and stone to feed them. The number of volunteers working was only outnumbered by their friends and family members who cheered them on as the concrete inched its way across the bridge. As the sun started to disappear behind the mountain ridge, a cheer rose up from the workers and observers that must have been heard for miles – the bridge was done!

The three beasts (mixers) are fed to complete the concrete deck.

A Celebration

The people were bursting to celebrate as the barrier that had held back the community's dreams for generations was broken. As the workers completed the final details on the bridge, meeting after meeting was held by the community to plan the celebration. Not a single detail would be overlooked. They could not control their excitement and a large party was planned.

The day finally arrived, and the mountainside came alive with the movement of people making their way to the party. It was not a good day to be a chicken in the community as countless birds were brought by each family to be prepared for the grand event. Speeches were given and at least one million "Gracias" were given as the people hugged each other in the excitement. A large, old oil barrel was filled with the local moonshine and refilled when it started to run low. It would be an understatement to say that "A good time was had by all."

Six months later, back in the United States, the bridge was nominated by the American Society of Civil Engineers for its Outstanding Civil Engineering Achievement Award given to the top projects around the globe. The project was selected as one of the six national finalists and the design team was invited to the black-tie affair in Washington, D.C. The design team felt a bit strange as they ate their steak dinners while wearing their best dresses and rented tuxedoes.

I overheard one of the spouses of a design engineer for the Golden Gate Bridge Seismic Retrofit project that had won two years earlier. "You are lucky the Guatemala Bridge was not up against you two years ago," she said. "I certainly would have voted for it over your project."

Sadly, the project did not win and as the design team cheered the winner, a groundwater replenishment system, James asked, "I wonder what Rolando is doing tonight? It is a shame he is not here."

"I have a feeling he is already planning the potable water project," I said with a smile.

Don Mike's Landcruiser
Photo Credit Scott Mitchell

CHAPTER TEN
The Landcruiser

*"The biggest risk in this country is getting
into an auto accident."*
Michael Shawcross

The Landcruiser was Don Mike's calling card, signature, and trusted companion all wrapped into one. It had been born in 1983 and spent its early years working on paved roads for another nonprofit organization near Antigua. When the organization moved back to England, they asked if they might store the Landcruiser in front of Don Mike's home. He gladly agreed. More than a year later, he contacted the organization who informed him that they would not be returning to Guatemala and offered to donate the Landcruiser to him. After another year of paperwork, Don Mike was able to secure the title and Don Mike and the Landcruiser were officially "married."

The truck had a cream-colored exterior with a boxy, no frills look. The mud and dust were easily seen against the light finish and the vehicle always seemed to need a wash. The interior was light brown which did a much better job of hiding the mud and dirt that inevitably made its way inside. The radio was a non-working relic of long ago and the truck was parked in the sun one too many times, which resulted in the windshield having a series of bubbles in the glass that permanently interrupting the driver's view.

Inside the cab it always had a faint smell of gasoline as the spare five-gallon gas can always seemed to struggle to have the cap fit correctly. The engine was a powerful V8 which never purred and always seemed to be missing the firing of at least one cylinder. The engine was always thirsty demanding a gallon of gas for every six miles traveled on the country roads. This would improve to an amazing eight miles to the gallon on the highway.

Occasionally, the Landcruiser also served as a makeshift hotel. The rural roadways of the Highlands were always unpredictable. It was not unusual to be stopped on the road when a large truck blocked passage due to mechanical problems. Since any alternate route might add days to the journey, Don Mike would tip back his seat and nestle into his "bed", waiting till the part might arrive the next day.

One time, an entire bus had lost its footing on the edge of the road and slowly rolled onto its side, blocking passage. We kicked back our seats and settled in for a long, restless night. As first light arrived, we were awakened by a gathering group of people. During such events, it was understood that everyone would provide whatever assistance they could, and we joined the gathering crowd. Soon, over 100 people using ropes and poles made from saplings rolled the bus back onto its wheels as the sun started to flood the valley with its rays. The bus then proceeded on its way with a honk of its horn and a wave from its driver and passengers.

When it was needed, the Landcruiser had power. The low gear of the four-wheel drive gave it the torque to climb the near vertical terrain that Don Mike challenged it with from time to time. It would grunt and groan as it pulled itself and its passengers up the path but always seemed to accomplish its mission. The vehicle's massive weight gave it traction as its four large tires worked in unison to pull the truck forward. The Landcruiser also had heavy skid plates protecting its underside and Don Mike was not afraid to use them as he bounced over the sharp rocks that were in his path.

The sharp rocks took their toll on the Landcruiser's tires. Changing a flat tire was a normal, daily activity and two spares were always at the ready. Don Mike hated buying anything new and always insisted on fixing the flat at the nearest *pinchazo*.

Pinchazos, or tire repair shops, are everywhere in Guatemala, often marked by a half-buried tire planted at the roads edge. They wait patiently to patch and repair the many tires that are damaged by the rough, unpaved roads. One time, after a particularly difficult trip, both rear tires were flat due to punctures from the sharp rocks. When I suggested that it might be time to purchase two new tires, Don Mike only scoffed and snorted.

"Those tires have a lot of miles left on them." It was only after I showed him the steel belts coming through that he reluctantly agreed to upgrade to a better, used set of tires.

In the steep mountainous terrain, the vehicle's brakes were its most important feature. On one trip, a sharp rock expertly reached up and ripped the brake lines off the truck. Soon the brake fluid was spilled onto the ground and the brakes were useless. After several failed attempts to work a temporary solution, we started down the mountain to find a mechanic.

The Landcruiser was put into the lowest gear of four-wheel drive and it slowly ground its way down the mountain. A few strange looks were given as people would pass the truck on foot but slow and steady was the motto. After coasting into the mechanic's shop after driving all night and using the curb to stop, Don Mike muttered, "That wasn't too bad."

"I'm afraid I have to differ," I replied as I uncurled my white knuckles from the steering wheel.

Sometimes Don Mike forgot that the Landcruiser was not a pickup and used it to transport almost anything. It was used to haul sand and gravel, water pipe, cement and lumber for the construction projects. I laughed out loud when I read the note on the luggage rack, "Maximum load 200 pounds" as I looked at the 1,500 pounds of cement perched on the roof. The abuse to the roof eventually took its toll and it sagged more and more as time moved on. Don Mike refused to acknowledge the fact until the water pooled on the roof to the point that it would drip down on top of his head, soaking his British flat cap.

He took the Landcruiser to his trusted mechanic who studied the situation. He scolded Don Mike for abusing the roof so badly and determined the only option to fix the vehicle was to cut off its roof with a torch like removing a lid from a tin can. He would then need to weld a roof back on from a Landcruiser found in the junk yard. It took three days to cut off the roofs from both vehicles and another three days to weld the new roof onto Don Mike's truck. After a weak attempt to match the truck's color with paint, Don Mike stood back and admired the work.

"Perfect," he exclaimed. "The doors even work."

Over the years, Don Mike and the Landcruiser were an inseparable team. The truck took Don Mike to where roads did not exist, using riverbeds and footpaths for the route. Other than the shoddy tires, Don Mike always made sure the Landcruiser was well maintained, checking all the engine fluids several times a day. He loved that truck.

"She's never failed me and complains much less than you do," he said to me as he patted the dashboard.

His mechanic in Antigua was almost like family due to the frequent visits. "Ah, you have brought her back to me," the mechanic would say as Don Mike pulled the Landcruiser into the shop. "What have you done to her this time?"

"She just needs a little attention," Don Mike would reply as he pulled out a long list of repairs needed.

The Highlands people came to recognize the cream colored Landcruiser as it was the only one of its kind in the region. As it drove down a mountain road, people would shout out "Don Mike" with a hearty wave, with Don Mike always waving back with a smile and greeting.

Don Mike insisted on picking up travelers as he made his way along the roads. He would stop and offer a ride to those carrying heavy loads and insisted that there was always room for one more. The time with these travelers was not wasted, as Don Mike always struck up a conversation with the riders. As many of the trips took hours, there was plenty of time to discuss the current events and communities' challenges. He always paid particular attention to the children, pressing them about their schoolwork and their dreams.

Don Mike was always passionate about education, especially for girls. He had seen the difference educated women would make in a community and was a firm believer in education as a change agent. Many a trip was spent with Don Mike talking about the value of educating girls and pressing fidgeting parents on the importance of educating their daughters, who sat silently listening to the conversation.

On one occasion, a young lady called out to the Landcruiser. The truck came to a halt as she approached and shook Don Mike's hand. "Do you remember me?" she asked Don Mike.

"I'm sorry. I'm afraid I don't," he said.

"I was the first girl to attend the new school you built here nearly twenty years ago." She then went on to explain that she had graduated and became a teacher. Now she was the director of the school.

Don Mike beamed and promised to visit her and the students soon.

"That must make all of this effort feel like it is worth it," I quietly said as we pulled away from the young lady.

"Yep, every time I start to think that this might be a waste of time, something like this happens," Don Mike replied as he looked thoughtfully out at the beautiful valley.

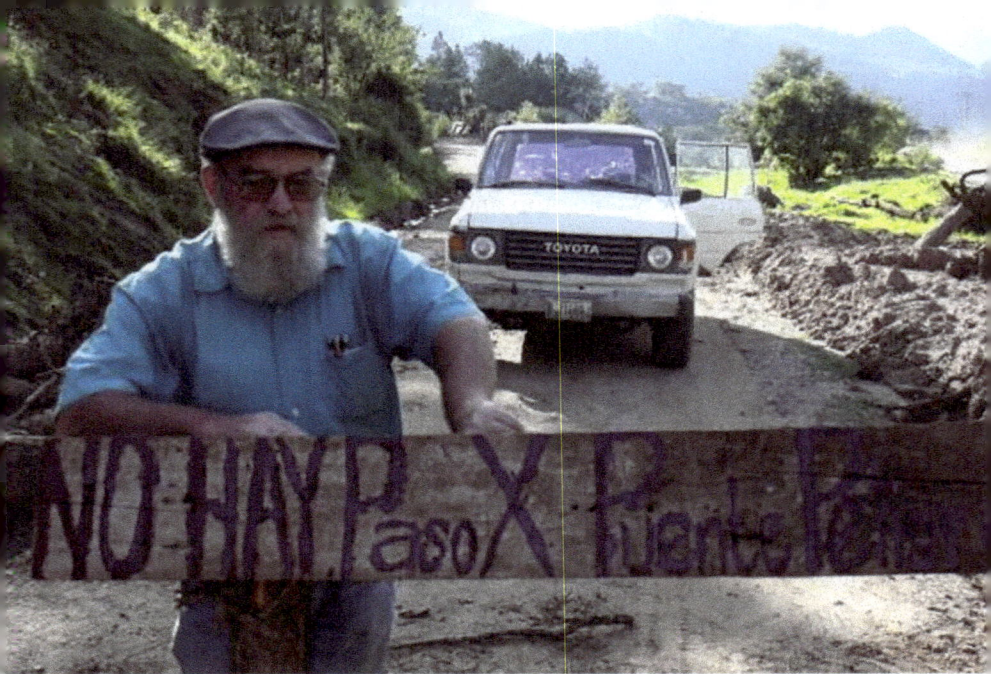

Don Mike poses in front of a 'Bridge Out'
sign after driving the closed road in his
beloved Landcruiser

CHAPTER ELEVEN
The Request for Water – "Water is Life"

"'Water is Life' and as such,
water rights need to be thoroughly investigated."
Michael Shawcross

After the bridge construction, the community was tired and exhausted. The effort had taken a tremendous amount of energy and a total commitment by everyone. The focus and dedication to the work had caused them to partially neglect their fields and other duties that everyday life demanded. After the construction, it had taken some time to catch their breath, recover and rest.

It was a year later when Rolando made the trip back to Antigua to visit Don Mike. The roadways to the bridge were now completed and vehicles and people could safely pass over the Rio Motagua, which was thankfully at risk of losing its nickname of "The Assassin". The river was no longer a barrier to the new markets and many farmers were beginning to grow more lucrative cash crops such as tomatoes and broccoli. Families were now presented with education options beyond primary school as the whole world of higher education was now within their grasp.

Rolando's mother, Gavina, was overjoyed as she now could send her patients to the hospital when needed. This was especially important for the expecting mothers to receive the expert care when she knew something was not right with the pregnancy. She no longer had to spend the sleepless nights worrying about a problem pregnancy and the frustration that her patients could not access the hospital.

But she reminded her son that the real improvement in the community's health would come with safe drinking water. As one of the community's health care providers (midwife), she knew how to improve the health of the people. But it was impossible to do it without access to clean water.

She was frustrated by the fact that two to three babies died each year due to the infections that resulted from a lack of clean water. On average, once a year she would lose one of her friends in the process of childbirth, leaving a family to fend for itself without its mother. Every time she would visit the family after they lost their mother, she could not help but wonder how their lives could have been different with a mother's support and love. She shuddered to think how her own family might have fared if she was not there to love and nurture them.

She was also concerned about herself and the other midwives. On a daily basis, they were exposed to the disease and infections of their patients without the ability to properly clean themselves or the patients they were committed to help. The patients, who were their friends and neighbors, needed them and they did everything they could do to help, despite the risks. Time and again, one of the midwives would get an infection from a patient – an infection that could be easily prevented if they could simply have access to clean water.

"As the community's health care provider, how can you attend to your patients' needs when you are sick?" she lamented to Rolando.

She had a vision and knew that clean water was the base upon which to build a healthy community. It would allow food to be properly washed before preparation. A reliable supply of clean water would also allow people to wash their hands and themselves on a regular basis, preventing infection and pathogens from entering their bodies. She shook her head when she watched the school director instruct his class on how to properly wash their hands – asking them to imagine that they had clean water to use.

She knew the school director was also frustrated. "How can I educate the young minds of the community when they were not in school due to diarrhea?" he would exclaim to her.

Each student was missing two days a month due to the illness and it continuously interrupted his lesson plans as the students were constantly trying to catch up because they had missed classes. Even when the students were in class, they were many times too ill to concentrate on their studies.

Many people did boil some of their water knowing it was contaminated, but it required a tremendous amount of firewood. Firewood that was in short supply and more difficult to find each year. Many families assigned two of their children to fetch firewood and water from the river, which restricted their ability to attend school. Most of the time, the daughters were selected to perform these tasks.

"How does a mother choose which child will go to school and which must stay home and collect firewood and water from the river? It is a choice that no parent should be forced to make," she told Rolando.

As she thought of the community's young ladies wasting their time gathering contaminated water from the river and firewood to boil it, she got angrier and angrier. These girls had so much potential and wanted to go to school, but it simply was not an option for them. How many of them could become doctors or other professionals, if they could simply attend school instead of fetching water and firewood?

She was determined to end this curse on her community.

Her voice was not going unheard by the community. Even before the bridge project, the community searched for a spring that would provide them reliable water that was also high enough to allow it to flow to the community by gravity. They had seen other communities saddled with the cost of buying diesel for a generator to power a pump in a drilled well and knew that it would challenge the sustainability of the system. The cost of diesel could take up to one third of their total income, a burden they simply could not afford.

Finally, a spring was found in a neighboring community, Xecoxol, about six kilometers from La Garrucha. The water committee visited the site with the owner of the land around the spring. Rolando remembers them gasping as they saw the water flow freely from underneath a beautiful banana tree and down the valley. It was the answer to their prayers. Rolando and the other members of the committee reached their hands into the cool, clean water and raised it to their lips. It was so refreshing, and they dreamed of the same water passing out of a tap at their homes – knowing the community would never be the same.

The cost of the spring rights would be Q30,000 or approximately $4,000. Rolando and the other water committee members worked tirelessly collecting the money from the homeowners.

Gavina continuously reminded everyone of the importance of clean water. "Yes, it will be a struggle, but buying the spring was well worth it to remove this barrier to their dreams," she said time and again.

Each family made the sacrifices necessary to raise the money by selling chickens, pigs or receiving remittances from family members working in the United States. The process took two years, but they were determined and never wavered.

But there was a problem they had not anticipated. The spring was in an adjacent municipality which passed a law that no potable water source could be used outside the municipal boundary. The law was passed after the purchase of the spring and they were hoping to be allowed to build the system since the spring had been purchased before the law had been passed. This was the reason for Rolando's visit to Don Mike.

As Rolando finished explaining the situation to Don Mike, he looked into the eyes of the bearded foreigner and asked if he could help.

Don Mike walking through a Maya corn field
Photo credit Scott Mitchell

Don Mike remembered stroking his beard and looking out the window at the volcano as he contemplated his answer. He understood the premise of the new law and the logic to protect the potable water sources within the municipality. Springs were relatively plentiful within the municipality on the south side of the river but communities outside the municipality with few options were repeatedly seeking to pipe the water to their homes. He understood and appreciated the municipality's desire to limit the practice.

He told Rolando that he knew the mayor of the municipality from several previous projects. He would like to visit the spring and then they could meet with the mayor to see what options might be available.

A week later, Don Mike, Rolando, several members of the La Garrucha water committee, and I bumped along the roadway towards the spring in Xecoxol in the Landcruiser. Although the spring was only six kilometers from La Garrucha as the crow flies, the winding, mountain roadway would take an hour for us to drive to reach the path to the spring.

I remember making our way down the steep path that wound its way among the trees until it opened into a beautiful meadow. On the far end of the meadow stood the large banana tree and the spring flowing out from under its base. The spring was strong and clear as it flowed freely from the earth. We took water tests that would confirm that the quality was suitable for drinking. I remember cupping my hands and tasting the water for the first time. It was so cold and refreshing after the hike and seemed to energize my entire body. It was no wonder that the people were working so hard to get this life changing water to their homes.

After visiting the spring, we returned to La Garrucha and stopped by the school for a Coca Cola and a tortilla snack. Don Mike was pleased as the grounds continued to be free from trash and the trash barrels were clearly being used on a regular basis, just as he had suggested years ago during the bridge project. Don Mike commended them for their efforts of educating the youngsters on the importance of keeping the land, which is such a wonderful gift, free from trash. They smiled back at him, proud of the accomplishment and their school.

He then noticed the stairs leading up to the school that had fallen into disrepair. Erosion had undermined the steps and they were at risk of falling down the slope. The water committee saw his disapproving look and explained that they knew the repairs were needed but the school lacked the funds for the materials.

Don Mike scoffed. "Surely there were sufficient resources within the community to make such a small repair. How could you ever be entrusted with a water system if you cannot keep the school steps in good condition?"

The water committee continued to push their case, but Don Mike would hear nothing of it.

Don Poncho was a member of the committee and the preacher at one of its churches. He also is an experienced mason who had built many of the community's buildings. He is middle aged, stocky and strong with muscles well developed from working his fields and construction. He has a well-kept beard and wears a wide cowboy hat to shield his head from the sun.

I noticed him quietly stepping away during the discussion and then return with a pail of cement, a shovel and a trowel. I remember that without saying a word, he began to do the repairs. The rest of the committee soon noticed and stopped their argument with Don Mike and joined in.

With the homework started, Don Mike now knew that the community was ready for its water system.

A week later, Rolando, the committee and Don Mike sat in the neighboring municipal offices. Rolando had asked for the meeting to seek approval to use the spring to provide water to his community, even though it was in a neighboring municipality.

After exchanging pleasantries, the Mayor explained the purpose of the law and how the municipality needed to manage their water sources for the benefit of its people. He wished he could share the water within his municipality with everyone, but he had to ensure that there would be enough water for his people first.

Rolando remembered his heart sinking as he knew there was logic to the law that prevented the water from leaving the municipal boundary. If he lived in the Mayor's municipality, he would applaud him for looking out for his constituents' needs first.

Don Mike then explained that the spring in Xecoxol had been purchased prior to the law being passed and Rolando handed over the bill of sale with the date clearly highlighted. The mayor reviewed the document and asked if the community of La Garrucha would be willing to sell the spring. The municipality would be willing to buy the spring so the community of La Garrucha could recover its investment.

Rolando then explained the extensive search that was done to find the spring. They had few other options and he begged that an exception be granted. Don Mike added his support and confirmed that the spring had been purchased prior to the law's passage and the community had few alternatives. The mayor looked at the desperate members of the water committee and agreed that they could proceed if they could secure the right of way for the pipeline.

Rolando remembered the relief he and the committee felt as they allowed themselves to smile again. He had feared that all the work they had done to raise the money for the spring would be for naught. Now they could proceed with the project. They thanked Don Mike for his support and asked if he could continue to help them with the project. He had grown to admire the community, especially Rolando, and agreed to help.

As the committee returned to the community, word spread rapidly. The water project was back on track! Rolando remembers Gavina running to him and giving him a hug and would not let go.

Rolando

Don Mike with students and their mothers.
Photo credit Scott Mitchell

CHAPTER TWELVE
Designing and Funding the Water Project

"A Water Study takes time and must consider the life cycle operations
and maintenance as well as the initial construction."
Michael Shawcross

The community had paid a pipe company $1,600 for a design that estimated the construction costs to be $150,000, Don Mike reported. The cost seemed extravagant and Don Mike was concerned that the pipe company had taken advantage of their conflict of interest and overdesigned the system to sell more pipe.

I discussed the project with Dr. Zitomer at Marquette University to see if their EWB chapter might be interested in the design and construction of the project. It would be a major project for the student chapter, requiring a year of engineering studies and up to another year for construction.

He told me that he had been approached by several of Marquette's best engineering students who had expressed interest in taking on a larger and demanding water project. He promised me once again that the project would have access to the "best of the best" students Marquette had to offer and that he would support the project if I was the team's mentor.

But we both knew that the design and construction was the easy part. We knew that a potable water project required continued support and coaching after construction to support the utility. EWB had recently completed a monitoring and evaluation study showing that water utilities without long-term social and technical support struggled to be sustainable. But with Don Mike committed to support the community, both of us felt comfortable moving forward.

Dr. Zitomer suggested that the study and design be part of a two-semester design elective. He added that it would be a good candidate for the US Environmental Protection Agency's (EPA) program called "People, Prosperity and the Planet" or P3 for short. Through the program, grants were awarded to university teams in two phases. Phase I was $10,000 to a university team to perform the study and design. Phase II was a grant of up to $75,000. It would be awarded for implementation to only a few teams with the best projects as judged in a competition in Washington, D.C.

A grant application was submitted using the design elective class at Marquette University. Just as the university was finalizing its fall semester classes, the good news arrived - they had been selected!

Four students enrolled in the class with me listed as the instructor. On the first day of class, I could see that the students were excited and eager as they sat on the edge of their chairs. Excitement filled the room and I could tell that I was to be blessed with another great group.

Amy was passionate about water. She was of modest height and had flowing blonde hair, but energy was her calling card. Her unending energy and smile made friends easily and she was more than willing to strike up a conversation with a stranger. Her enthusiasm and positive attitude picked up any group and carried them over any rough patches.

Jack was athletic, with a classic "tall, dark and handsome" look. He was extremely intelligent but reserved in his actions. Engineering and service had always been of interest to him and he was excited to have them combined on a common project. In addition, Jack spoke fluent Spanish and was excited to put his language and engineering skills to use on the project.

Alicia was looking for an interesting class to fill out her full-time student schedule. After hearing me speak at a presentation, she registered for the class because it sounded interesting and fulfilling. She was smart and organized, making her an important part of the team. Alicia loved animals and had an engaging laugh that one could not help but fall in love with.

Mark was tall, reserved and had an adventurous spirit. Having grown up on a goat farm in California, he also was comfortable around animals and understood an agricultural society. His small farm background has also given him the ability to see creative solutions to most any problem. He was extremely intelligent, which combined with his creative thinking, made him an important part of the team.

On the second day of class, a bright-eyed freshman entered the classroom. Adrianna had seen the class posting, but as a freshman, was not eligible to enroll in the class. She was hoping she might be able to audit the class, as the subject was exactly why she had chosen civil engineering as her field of study – to help others through mathematics and applied sciences. She fidgeted with her hands as I contemplated her request. I decided to ask the four senior students what they thought, and they were happy to add her to the class. Plus, Amy said. "We can really use some drafting help with the plans, and she knows Computer Aided Drafting (CAD)."

"Yah, we could really use a drafting monkey," Jack said. As I looked back into her hopeful brown eyes, how could I say no?

The class would include a site visit paid for by the grant to gather data and meet the community. The students were excited, and Amy pulled together check list after check list to ensure that all the equipment needed would be brought. The energy was high with anticipation as none of them had ever traveled to a developing country before, but I could sense that something was wrong with Adrianna. She looked worried and fretted as the rest of the students excitedly made plans. When I finally asked her if she was okay, she quietly asked if I would be willing to talk with her mom as she had concerns about her traveling.

When I called Adrianna's mom, she immediately dove into the conversation with emotion. She clearly articulated her concerns about her daughter's safety. For what seemed like the better part of an hour, she spoke firmly and with conviction without allowing me to respond with a single word. It was clear that her mind was made up and she would not be swayed by any information.

She wrapped up by saying, "I think taking students to Guatemala might be the worst idea I have ever heard of. Taking my daughter on this trip is simply not an option." The phone was silent for a minute until she asked, "Are you still there?"

I responded that of course, I would not take her daughter on the trip without her blessing and that her daughter would be safe. But this was a mother concerned for her child and she would not waiver.

I met Adrianna at the union and relayed the bad news to her over a cup of coffee. I could see her anger rise. "I'm 18 years old and old enough to make my own decisions," she exclaimed. "I am a grown woman, and she is not going to stop me."

I took in a deep breath and asked Adrianna to do the same. "I know you are upset and passionate about going, but your relationship with your mother is important, too. Don't jeopardize that over this trip. Please don't do something that you will regret the rest of your life."

I asked her to take some time over fall break to explain to her mom how important the concept of service engineering was to her. She was only a freshman and there would be plenty more opportunities to travel in the future. "I know you are angry, but please have some patience. It's only because your mom loves you so much that she is so concerned."

Reluctantly, she agreed as her anger subsided, but I could see that she was still very disappointed. Later, Adrianna would go on several international trips using her engineering skills and continues a life of service today.

I waited in the Milwaukee airport with the students as Amy read off her list and Jack, Mark and Alicia emptied the bags to ensure all the equipment was there. We arrived in Guatemala and made our way to Don Mike's home as the students took in the sights of a new country. The students were captivated by Don Mike's stories, character and British accent. They looked around his home full of books and hung on his every word as he told story after story about his life in Guatemala. I could see that Don Mike was in his element as he loved talking with young people who had such energy and optimism. I had learned over the years to give the students some time to take it all in and absorb this strange and beautiful country.

The students then met Katie, a Peace Corps volunteer assigned to work with the NGO PAVA and Don Mike. She was from Ohio and already had developed a passion for working with communities on development projects. She was a young, confident woman who had a rugged beauty about her with long dark hair and eyes that danced when she smiles. The students immediately connected with her as we bounced along the roadway in the Landcruiser, listening to Don Mike's stories.

We were greeted by Rolando who guided us to Don Poncho's church. The church was a large block building with fresh paint and a freshly groomed courtyard. The interior had a concrete floor, a beautiful altar and approximately fifty folding chairs for the parishioners. A car-battery-powered loudspeaker pointed out over the valley to announce the beginning of each service.

On top of the church, there was an odd, small room built of block that served as a storage area. We moved our bags to the room above the church where we would be staying. The bags were stacked in the corner of the small room to the ceiling to make space for our bed mats laid out on the floor. We then met Don Poncho's family who would be our gracious host during the stay. All seven children, four girls and three boys, lined up to greet us while Don Poncho's wife, Lucinda, fussed over a fresh pot of coffee and tortillas to refresh her guests.

I had brought some M&M chocolates for just such an occasion. As I opened the bag, each of the children shyly held their hands out for a few of the tasty treats. When I reached the end of the line, Ramone, the youngest boy, took off his baseball cap hoping that it might be filled with the candies. Everyone laughed and I had to admire his spunk, which I rewarded with a double measure of the chocolates.

We then met with the water committee in the church and developed a plan for gathering the data needed. It was the end of the rainy season so showers would be expected each afternoon during our stay, slowing the progress. We divided up into three groups.

One group would survey the route of the pipeline. The community had identified the route from the spring to the water tank site on a hill above the homes. The work was brutal as each step required the vegetation to be cut back with machetes. We slogged through the mud caused by the recent rains carefully taking our measurements. We also carefully laid out the location of two water line bridges that would be needed to cross rivers along the route. Each night we returned to our little room on top of the church wet, tired and covered in mud.

The second group investigated other water sources to see if a more cost-effective solution might be found. They reviewed the option of installing rainwater catchment systems on roofs, taking water from the rivers and piping water from a lagoon that was located high above the community. The lagoon was covered in a bog, or mat of floating vegetation that had grown on the surface over the years. We all laughed at them as they bounced on it like a trampoline, sampling the water for quality and depth. They returned to the church each night completely wet, but full of stories of bouncing on the bog and plenty of smiles on their faces.

The last group, led by Don Mike, visited the homes and carefully marked each location with GPS. The families were interviewed to collect information on what water source they currently used and any health problems that they were experiencing. Water samples were also taken if water was on site to determine its quality. As each mother was interviewed, the students could see her excitement build. Maybe her daughters would not have to fetch water like she, and so many other generations of women have had to do for their families. The group members soon found that their biggest challenge was trying to extract themselves from the family's hospitality that always included an offer for hot coffee and a tortilla.

Don Mike had been careful to explain to the students that Highlands dogs are not pets like in the United States. These were working dogs and the students should refrain from the urge to pet them. The last thing Don Mike wanted to deal with was a dog bite. After the first afternoon of hiking, Alicia returned to the church with two dogs in toe.

"It's not my fault," she complained to Don Mike as he shook his head while she scratched her new friends under the chin. "They came up to me and are so cute!" For the rest of the trip, she always seemed to have dogs, pigs or goats at her side.

Marquette University had also sent the editor from the alumni magazine, Joni, and Kat, a photographer to capture the story. Kat, the photographer was taken by the community and the assignment. She buzzed around taking photos of the team and the amazing landscape. She liked the outdoors and the rain and mud were not going to slow her down.

One evening, she explained to us that she was enjoying taking the action shots of the group and landscape views of the Highlands, but her specialty was portrait photography. Don Mike asked, "Would you consider taking family portraits of the community members?"

Kat looked back with surprise. "Do you think that they would be interested? Do you think it might be possible?"

Don Mike smiled back and nodded.

I had a feeling that something special was about to happen.

Don Mike had taught me that it is important to engage the community on many levels, not only the infrastructure project. As an engineer, I sometimes found myself lost in the technical details of the project and didn't recognize opportunities to engage the community in their everyday lives. Over the years, Don Mike had shown me that integrating into the people's lives built valuable trust and respect.

Word was sent out across the valley using Don Poncho's loudspeaker system that anyone who was interested in a portrait should arrive at the school on Saturday at noon. As the time approached, we rode up to the school in the Landcruiser. Kat was nervous. "What if nobody comes?" She said looking down at her camera.

As we rounded the corner to the school we were greeted with an amazing scene. Nearly every family was at the school, all dressed in their best clothes for a family portrait. Mothers and grandmothers were doing their best to keep the children's' clothes clean while the children ran around playing with their friends. Fathers and grandfathers stood smartly in their best clothing while their wives continued to fuss over them, brushing off any little imperfection.

Kat was in her element, setting up a makeshift studio with the proper lighting. She would give the families the best possible photo that she could provide. One by one, the families came forward for their portrait. Even though Kat had hundreds of photos to take, she refused to rush the process, taking time to make sure that every child was sitting "just right" and the photo was perfectly lit. She even recruited some helpers with toys to get the toddlers to smile for the photo at just the right time.

A battery-powered printer had been brought on the trip and it soon was printing out the portraits one by one. A crowd several people deep gathered around the printer awaiting the arrival of each photo. Although it took all afternoon, the crowd never tired of guessing whose portrait was coming out next. As soon as a family was recognized, a cheer would go out and the recipient would blush as they collected the photo, treating it like the prized possession that it was.

The community gathers around the printer waiting for the photo to emerge

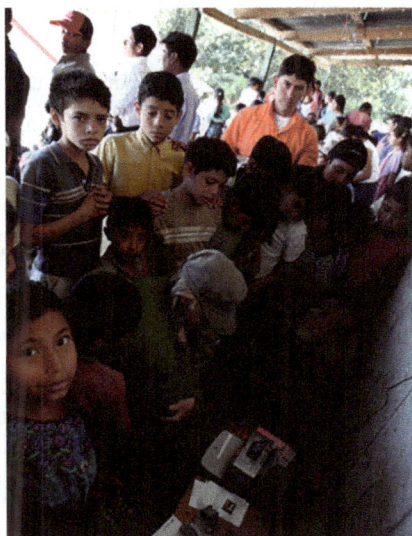

Years later, when I happened to visit a family's home, I usually saw their portrait proudly stuck above the table on the wall. In many cases, it was the only family photo they ever had.

We returned from our short trip to La Garrucha with a new purpose. Meeting the community and seeing how they lived made the project real to the students and they were determined to help. They poured themselves into the work with renewed energy. Alternatives were developed that used the Xecoxol spring water, rainwater catchment and water from the lagoon. Preliminary designs were developed along with construction and operational cost estimates, and a list of advantages and disadvantages that would be presented to the community.

We returned to Guatemala in January during Christmas break to continue our work on the project. As we left the cold Wisconsin winter behind, we looked forward to the warm sunny skies of the Highlands. The landscape was transformed from the mud and lush vegetation of the wet season to the brown and dust of the dry season. Things looked so different, but beautiful in their own way as we had forgotten just how amazing the landscape views were.

The students looked forward to reuniting with the community as we drove down the bumpy road in the Landcruiser with Don Mike and Katie. Don Mike updated us on his last meetings, and the students asked him questions about his adventures, which of course led to even more stories. One of their favorites was the one where he worked for two months to find a mule train to transport cement five hours along a mountain path to a remote bridge site – only to find out that the community had carried the 150 sacks of cement to the site on their own backs!

Once again, we were greeted heartily by Rolando and made our way to Don Poncho's house where we moved into the room above the church. Lucinda, Don Poncho's wife, had a fresh pot of coffee brewing and fresh tortillas on the fire and it all felt so comfortable.

Soon, Alicia was out in the yard reacquainting herself with her furry fan club of goats and dogs.

During the previous visit, rainwater was harvested from the roofs of the church and Don Poncho's house to provide ample water for the team's use. On this return trip, since it had been months since the last measurable rain, the family was using water sparingly as Don Poncho's daughters fetched water from other sources.

The Team's women, Amy, Alicia and Katie, expressed frustration that the fetching of the water was the responsibility of the girls and not shared by the boys.

Don Poncho looked surprised and tried to explain, "Fetching water and firewood is always the job of the girls. It has been that way for generations. It is the natural order of things that the men worked in the field and the women work in the home. Who else would fetch the firewood and water needed to support the house?"

The Team's women were not impressed, and Don Poncho's daughters listened intently to the debate. It was unclear to me if any minds were changed but it was clear that the discussion got people thinking.

The students updated the water committee on the study. They discussed the system options, test results, advantages and disadvantages of each in detail.

Rainwater catchment was proving to be very difficult given the six-month dry season. This solution would require very large storage tanks for each home to store six-months of water supply for the family. The storage tank required would be larger than the home itself in many cases, which would not be realistic.

Using the lagoon as the potable water source was more promising. The biggest advantage of using the lagoon water was that the initial construction cost was less due to the shorter conduction line, since it was three kilometers closer to the community than the Xecoxol spring. The biggest disadvantage was the operational costs of the system as the lagoon water would need to be treated to remove the bacterial contamination. Don Mike had taught me that even the simplest treatment system could be a major barrier to a system's sustainability.

Using the Xecoxol spring as the potable water source would have higher initial construction costs when compared to the lagoon, but this option would have lower operational costs, as complex water treatment would not be needed. It would be easier to maintain and have lower monthly fees for the users.

After some discussion, the committee directed the students to proceed with the Xecoxol spring water source design as they believed they would be able to obtain the right of way needed for the pipeline. The students worked with the committee to gather additional data on where the pipeline route needed to be changed due to uncooperative property owners. The surveying was much easier without the mud, and the machetes sung as they blazed through the brush. We also collected an additional round of water samples to check for any differences between the rainy and dry seasons. I had learned from Don Mike that sometimes the water quality results would vary from season to season, due to agriculture practices, such as contamination due to pesticides.

We did not have access to an incubator for the samples which is typically used to allow bacteria in the water to grow and be observed. We chose instead to place the water sample tests in a sock and tape it to our bodies to keep them at the correct temperature. It was quite the site as we crawled into our sleeping bags at night with socks taped to our bodies like large leeches to keep the samples warm.

"These socks look like a strange new fashion statement," Amy said as she surveyed her teammates.

After the samples were incubated 24-hours, they could be read by holding the sample up to the sun and counting any bacteria colonies that might have grown.

On the last day, we said our goodbyes and started driving away when Mark noticed that one of his socks filled with the previous day's water samples was missing. A frantic search ensued that resulted in the Landcruiser being emptied along the side of the road. Still, the missing sock could not be found.

Don Mike then received a call from Rolando. A member of the community had found a strange item in the middle of the field and given it to him – it was the sock!

By this time, Rolando was adept in reading the sample results so he taped the sock to his own body and would report the results the next day when the samples were ready to be read. A sigh of relief went up from the students and they repacked the Landcruiser.

As we jostled along the roadway, the rear door of the Landcruiser flew open after an especially large bump. It must have not been fully latched after the repacking efforts and I screeched the Landcruiser to a stop. The Team's luggage lay spread across the roadway and ditch. Quickly, we scrambled to recover the items and reload the Landcruiser with each person making sure their bags were recovered. Several hours later, when we arrived at Don Mike's home, I asked if anyone had seen the turbidity meter.

A turbidity meter measures the clarity of a water sample, which is important information when determining water treatment options. I had borrowed the device from my employer for the trip, promising to return the $1,200 unit safe and sound upon my return. The lab manager had said that borrowing equipment for such a purpose was not within the company's policies, but after hearing my story about the community and the student's project, he reluctantly agreed to release the meter. "Just don't lose it, or we will both be in big trouble," he said as I walked out the door.

After a frantic search, it was clear that the meter was missing. We knew the most likely explanation was that the meter was still alongside the roadway, having fallen out when the Landcruiser door fell open. I felt sick as I thought about explaining the lost meter to the lab manager and my boss. It was a two-hour drive back to the spot where the luggage had been spilled and the team's flight was departing in six hours. The time frame was tight, but I felt I had to go back and do everything I could to find the meter.

"At least I can tell them that we did everything we could to find the meter," I told Don Mike.

As Don Mike and I roared back in the Landcruiser to the site where the meter was likely lost, I knew the chances of success were slim to none. We had done a thorough search before and the meter could be anywhere. Miraculously, we found the exact spot where the luggage had been spread across the roadway and diligently searched the ditches and roadside brush, slashing the brush with Don Mike's machete. No trace of the meter could be found.

"Well, it certainly was a long shot," I said to Don Mike.

A hundred meters up the road, we noticed a bus stop with a young man sitting on the bench waiting for a ride. We walked up to him and asked if he might have seen a blue suitcase lying alongside the roadway, as it was ours and had fallen out of the Landcruiser a few hours earlier.

To our amazement, the young man said. "Yes, indeed. Was it a light navy-blue suitcase with wheels?"

It turned out that one of the homeowners nearby had spotted the suitcase and upon opening it, had no idea what the device inside was. He had asked the young man, who was a photographer, if he might know what the expensive looking device might be. After the young man had looked at the device, he gave it back to the homeowner saying he had no idea what such a device might be used for. The homeowner took the suitcase back to his house.

My hopes rose and said to Don Mike. "Well, this just might be my lucky day."

We then knocked at the homeowner's door and a young man answered. We asked him if he had seen the blue suitcase and explained how it had fallen from their vehicle a few hours earlier. The young man did not immediately answer the question, but after some discussion, it became clear that he knew where the suitcase was. He then asked for $200 to return the meter, saying that his brother had given strict instructions to take nothing less if the owners of the meter returned.

Don Mike exploded into a tirade of swear words and insults. "How can you ask for payment for the meter when it is not even yours?" He thundered. "Don't you know that the meter is being used for a volunteer project that will provide potable water to a community? Don't you know how important these projects are to the people?" Then he said threateningly, "Should I go get the community to come and explain how important this project is to them?"

The young man seemed unshaken and still demanded the $200 saying, "Sorry, it's not my decision. I am only following the instructions from my brother."

Don Mike had seen plenty of scams in his life and was determined to not be taken by this one. He kept arguing his points with the young man and raising his voice.

I kept checking my watch, knowing the flight time was fast approaching. I pulled Don Mike aside and said, "We have been incredibly lucky that we have found the meter at all. Let's rejoice in our good fortune and move on."

Much to Don Mike's disgust, I handed over the $200 to the young man, who then retrieved the blue suitcase with the meter inside, safe and sound. It would take years before Don Mike forgave me for paying the "finder's fee".

When the students returned home, they began the final push to complete the design. They met nearly daily, having "their spot" claimed in the student union. Twice a week, I joined them to answer questions and provide encouragement. I knew their other studies lapsed as they threw themselves into the project. Clearly, the class was more than a grade. It was a calling to help people in need.

The Team worked to find creative solutions to reduce costs without compromising quality. In the spirit that no chain is better than its weakest link, the team sought out any component that was over designed and redesigned it to reduce the cost. The Team also brainstormed creative solutions to reduce the cost of the system. Over the years, I learned that food was an excellent motivator for students, and so I offered to take them out to dinner for each creative idea that they came up with. Idea after idea was submitted to me for evaluation as the students got more and more creative. After the ideas were reviewed to ensure that they could be constructed in Guatemala by the local people, I would "pay up". Hence ideas that received the label "dinner line" and "dinner tank" were born, reflecting the victory of receiving a free meal.

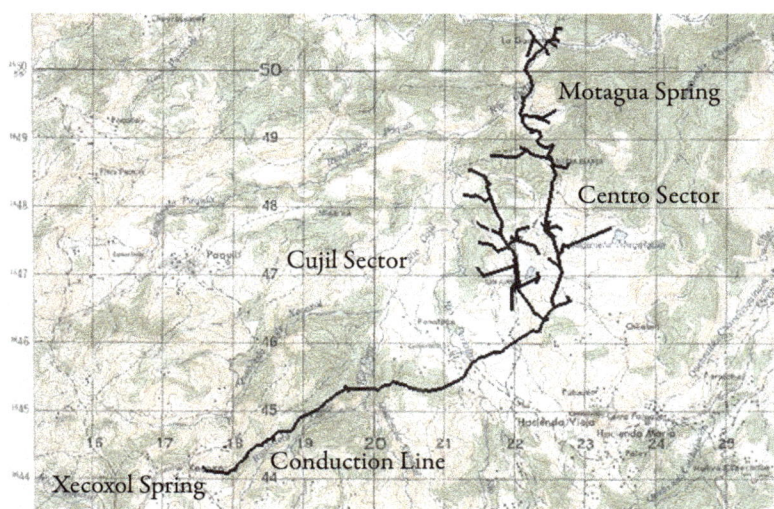

Map of La Garrucha potable water system

As the Team ate their corn tortillas and chicken soup at the Antigua Real Guatemalan Restaurant outside of Milwaukee, they felt proud that they had reduced the cost of the project from the initial $150,000 estimate. After contributions from the municipality and the community, the funds needed by EWB for materials was $75,000 – which happened to be the amount of the EPA P3 second phase grant – if they were successful. Their focus was now turned to the grant application and presentation that would be made in Washington, D.C. on Earth Day. They owed it to their friends in the community to do everything they could to win.

They endlessly fussed over the graphics that would be used for the display and practiced their presentation until it was well polished. When the time came for us to fly to Washington, D.C. for the National Student Design Expo, they were ready and confident that they had done everything they could do to win. We went directly to the capital's mall and set up our tables in the assigned spot.

Several large tents had been set up on the mall for the competition. It was springtime and the cherry blossoms were in bloom, providing a lovely fragrance in the air. The typical distractions of Washington, D.C. were everywhere, but the students remained focused on their task, hardly noticing their environment.

After setting up their display, they then reviewed the other fifty teams' projects and their hearts sank. The competition was going to be stiff and they found out that there would be only six teams selected for the Phase II grant award.

"There are some really smart people here," Jack said with concern in his voice and they all agreed.

The students were able to meet Wisconsin Congresswoman Gwen Moore and Wisconsin Senator Russ Feingold in their offices, a first for all of them. Other members of Congress walked by the booth and could not help but to engage the students about their project. Soon there was a large group standing next to their booth as the students explained their design, answered questions and told stories of their travels to the site. Their passion and commitment were clear to anyone who happened to walk by, and those passing by could not help but be impressed with the young engineers.

The time came for them to give their presentation and they were ready. Unlike other teams, the students would do the presentation themselves. I was so proud of them as they delivered the presentation professionally but with the passion and energy that only students can bring. They nailed the responses to each of the questions asked by the judges and soon the timekeeper needed to move the judges onto the next group. As one of the judges stepped away, I overheard him say,

"Now they were impressive."

Finally, the Team sat in the auditorium awaiting the results of the competition. They knew they had done their best, but was it good enough? They also knew that they had put all of their time and energy into the competition. It was an "all-in approach" as they had little time to spend on other funding options.

After the host made the introductions, it was time to announce the winners and the team leaned forward in their chairs. He then said he had a surprise. There would be one team that would receive an "Honorable Mention Award." Although the Team would not receive a Phase II grant, the judges clearly wanted to recognize their hard work and dedication to their project. He then announced the Honorable Mention Award would be given to "Marquette University – La Garrucha Water Design."

The students looked at each other horrified. They knew at that point that they had not been selected for a phase II grant and their hearts were broken. But they still needed to go up to the stage and receive their certificate. They slowly made their way up to the stage to the applause of the audience. They were clearly the favorite of many of the teams and a standing ovation was given to them as they forced weak smiles and shook the hand of the host.

Tears ran down the cheeks of Amy and the host mistook them for tears of happiness as he smiled and handed her the certificate. The feeling of disappointment was overwhelming as she felt that she had let down her friends in La Garrucha. She somehow forced a weak smile and said "Thank you."

As they flew back to Milwaukee in silence, the Team could not help but be disappointed. Now what would they do? How could they raise the money needed to make the project a reality?

Then Jack stepped up. "We have simply come too far to give up now," he said. "We just need to have a little faith and redouble our fundraising efforts. No one had said that it would be easy, and failure is not an option. Too many people are counting on us."

As the Team sat in the student union brainstorming on fundraising ideas, they called Don Mike and shared the bad news. He suggested that the $10,000 raised from the connection fee be used for the project's materials. He and the community could also press the mayor to increase his participation, asking him to pay the salaries of Mincho and the other masons who would work on the project. With these changes, EWB would still need to raise $55,000. I knew this was going to be a challenge, but had hope that with our growing list of donors, it might be possible.

The students started writing unsolicited grants, knowing that the chances of selection were low based upon our past experience with the bridge. But they had to try. They had limited opportunities as the university's Development Office would not allow them to contact anyone on its list of potential donors directly.

Our usual supporters had all agreed to help, but it was clear that we would need a new major donor if the project was going to be funded. We had never received a donation in this amount before, so I went back to my old friend Yash, and asked for his advice. Yash suggested that the Rotary clubs who had helped with the bridge would likely help with the water system as well. Water is a key area of focus for Rotarians and the project would surely be of interest. They liked the bridge project team and they would love to hear a presentation from the student water design team.

Being a Rotarian, he was also well connected within the community. He had a contact at the Helen Bader Foundation, which he thought might be interested in the project. He would set up a meeting.

When I met with the Helen Bader Foundation, they encouraged us to apply. They had heard the project's story from Yash and he had done his best to put in a good word for the Team. They liked the idea of partnerships - with the community and Rotary. They also liked that with a grant of $50,000, the entire, larger project could move forward. The students labored over the grant proposal, sweating each and every word. It was their last chance. Soon after the grant application was submitted, the Team received the good news that they were successful!

The Team celebrated with Guatemalan coffee in the basement of the student union. The Marquette EWB chapter was selling the coffee as a fundraiser to help pay their travel costs and it seemed only fitting that a toast using the "pride of Guatemala" was their beverage of choice. The project was really going to happen.

Rolando, Mincho, and the MU Design Team with Don Poncho's family

CHAPTER THIRTEEN
Setting up a Water Utility

"Clean water isn't free."
Michael Shawcross

Excitement built within the community as it eagerly awaited the start of construction. But Don Mike knew that there were some important decisions that needed to be made first and he called for a meeting with the water committee to discuss the utility's structure. The students were still at class at the university in Wisconsin, so in advance of the important meeting, he explained to them over the phone that building the water system was the easy part. Building the water *utility* was much more difficult.

This important part of any water project had to be carefully planned. He went on, teaching them that it was important to discuss the utility's operations before the first shovel went into the ground to give the community time to discuss and make the important necessary decisions.

"If these decisions can't be agreed upon with the promise of future water, they may never be fully addressed by the community and the utility will languish," Don Mike explained.

"I thought we were designing a potable water system. What is involved in a water utility?" asked Adrianna.

Don Mike explained that the water utility entailed all the elements needed to operate, maintain and sustain the water supply to the users. In the end, Don Mike explained that it all came down to the answers to three simple questions:

How do you generate the water bill?

How do you collect the water bill?

What do you do when someone does not pay their water bill?

These three simple questions are the same for water systems around the world, but they demanded complex and difficult answers. The answers needed to come from the community itself, not him or the students.

"It is critical that everyone in the community understand these important questions and support the decision in the end," he said. "What might work for one community may not work for another and this needs to be respected."

In the end, it was the community's decision on the final answers.

The water committee chose to meet in the church. They arranged the chairs in a circle, and I tacked up a large map of the community on the wall. (Of course, an engineer could not have a meeting without a map). Rolando chaired the meeting with Don Mike seated to his right. Don Mike suggested that the meeting be used to discuss the important decisions that needed to be made to operate, maintain and sustain the water system and he apologized in advance for all the details. It would be a long meeting and asked for their patience. He suggested that after their discussion, the committee meet with the entire community several times to discuss the important decisions.

"It is very important that everyone fully understand the options and have an opportunity to ask questions," he said.

The community meetings would be led by the water committee and Don Mike offered to be on hand to answer questions and coach them through the process. He offered to give examples of what had worked and had not worked on projects in neighboring communities.

He then said, "The most important point is that the utility must operate in a transparent and open manner to eliminate any concerns over mismanagement and build trust within the community."

He had seen suspicion undermine water committees before and it generally led to the failure of the entire system.

"The rumor mill is a well-oiled machine all over the world," Don Mike later said to me.

How To Generate a Bill and Determine Its Amount

The first topic was the need for maintenance. Don Mike explained that even when a system is brand new, some maintenance will be needed. Landslides occur, especially during the rainy season, and other accidents like a horse stepping on a pipe can lead to leaks. Most communities employ a plumber to walk the line on a weekly basis looking for leaks and locations that needed attention.

"'A stitch in time saves nine' is the old saying for mending clothing, but it also applies to water lines," Don Mike said. "With time, the work of the plumber will likely increase as valves and other fittings start to wear out and need to be repaired in a timely fashion. It is important to get off to a good start with the maintenance program from the very beginning."

Whether the plumber would be a paid or a volunteer position was discussed as the committee members carefully took notes in their notebooks. The water committee talked about how a paid plumber would require a higher water tariff, but a paid plumber would likely be more accountable and consistent over time. Don Mike offered that some communities started collecting the water tariff before construction began so that the plumber could participate in the construction and become familiar with its design and operations.

Don Mike suggested, "The decision is yours if the plumber is paid or a volunteer and there are good arguments for both options. But I highly recommend that you select the plumber before the construction begins so as much learning can occur during construction as possible from Mincho and his team of builders."

How the maintenance materials would be paid for was then discussed. The members of the committee were clearly engaged as they bantered their ideas back and forth. Don Mike offered that, generally, communities used one of two options.

Option one was to collect a small monthly fee that would include minor materials with the understanding that any major repairs would require a special assessment on the users. This posed some problems as the funds would not be available immediately when the repairs were needed, and the users might go without water for several days while the funds were collected.

The other option would be to collect a little extra money each month from each user and use the funds to build up a reserve. This would require a bank account in town to hold the funds and a community that trusted that the funds were managed properly.

"When it comes to money in accounts, people are always suspicious," lamented one member.

I then remembered that many organizations suggest that a larger water bill be charged to build up the reserve. Some even ask that the reserve be sufficient to rebuild the entire system after seven years.

"That certainly sounds good, until life gets in the way," Don Mike said as I told him of the policy of these organizations. His experience had shown that when the community was challenged with events such as a disaster or a poor crop harvest, the reserve fund was reallocated to other more pressing needs. I knew that even in my hometown in Wisconsin that was the case. The city raided the water fund account once the reserve was built up and the city determined it needed funds for another project. It seems that there is always a community project that is a higher priority than a reserve fund, no matter what country you lived in.

The next topic was how much of a connection fee should be charged. Don Mike offered that most communities charged a connection fee to each family that received a tap and also a monthly usage fee. In his experience, communities that had a small connection fee had been overwhelmed by requests for additional taps since they were nearly free. Families tended to request taps for their young children for when they would grow up and have their own families.

Other families requested taps for children who were not even born yet. He told of one community of 100 homes that requested a total of 500 taps for their future generations. This brought a chuckle from the meeting attendees. Don Mike went on offering that many times, the connection fee paid for the tap itself. This gave the user something tangible that they owned, and since it was theirs, they were much more likely to maintain it.

The fee to connect to the system in the future also needed to be established. This would help families to decide if they paid for taps for their children now or wait until later as they could weigh the immediate and future costs. The community also needed to decide on the connection fee for a families that moved from outside the community and had not helped with the cost of construction. Don Mike suggested that most other systems set this fee based on the funds and effort expended by the community on the initial construction to be fair to the initial users.

The room was now getting warm and I struggled to stay focused on the topics at hand. But I noticed that the committee members were taking the discussion seriously and writing detailed notes. One asked if he could borrow my pocketknife so he could sharpen his pencil, and soon the knife was passed around the circle as pencil shavings spread across the floor.

The final topic on "how to generate a bill" was the monthly fee. Two options were discussed. One was a flat monthly fee with the other being a fee based upon usage. A fee based upon usage would require water meters that would be read on a monthly basis and the fee calculated based on the amount of water used.

The fee based upon usage was considered "fair" in that those who used more water paid a higher fee, but it would require more work and bookkeeping. Another advantage is that it provided a solution for the community's most vulnerable members. Those on limited incomes could conserve water and had the option to use the river to do laundry to reduce their cost.

Don Mike offered that some communities initially used water meters to ensure fairness. They would publicly post the amount of water used by each tap to identify anyone who might be abusing the system by watering fields or livestock. Later when more users were on the system and water became scarce, meters could encourage conservation by charging a fee for the amount of water used.

I could see that the committee's heads were swimming with so many decisions needed, and so was mine. They decided to break and reconvene the next day to discuss the remaining topics. As they broke up and the committee members walked up the hill towards home, the discussions continued over the pros and cons for each decision. I could tell that they were fully engaged and committed to the system's sustainability.

How to Collect the Fee

The next day we reconvened at the church to continue the discussion. Each person sat in the same chair as the day before. The pencils had been freshly sharpened outside the church and the shavings blew along the ground seemingly chasing the chickens. The committee was eager to continue to discuss the decisions that would need to be made before construction could start.

The next item of discussion was how to collect the water tariff?

"That's easy," said one of the committee members. "That is one of the duties of the plumber, right?"

But Don Mike said, "The real issue is the need for transparent and open record keeping. Based upon my experience with other communities, once the users lose confidence in the management of the system's funds, they would refuse to pay the fee."

He suggested that many successful water committees appointed a treasurer whose job was to collect the fee and keep the records of who had paid and who had not. Some had even gone to the practice of posting the list of the users on a large piece of paper in a public place, showing the payments made and who was in arrears. He told the story of one community who posted such a list and a person was so embarrassed to be on the list of those in arrears, that he paid his bill on the spot and insisted that the list be immediately rewritten without his name.

In a similar manner, the expenses should also be tracked transparently. Each purchase should be listed, no matter how small. The records need to be accurate and current. Any user should be able to review the ledger at any time so that they would have confidence that the system was being managed correctly.

Finally, any reserve funds not only needed to be tracked but kept in a safe location. Of course, a bank account was preferred, but its use must be balanced with the ability to access the bank. Keeping large amounts of funds in someone's home should be avoided if possible. Mobile electronic money technology was not available at that time and has since provided much better options for water utilities.

What To Do if Someone Does Not Pay

When the question of what to do if someone does not pay their bill was asked, one member of the committee responded, "That's simple. We would turn off the water service to the user."

Don Mike responded. "So who, specifically would turn off the water?"

Everyone looked at the floor realizing how difficult it would be for them to turn off the water to a friend or neighbor. Or even worse, a family member.

"Who could ever turn off the water to their grandmother's home?" I thought, trying to put myself in that situation.

Don Mike explained the importance of handling the situation with compassion and fairness. "Who among us has not been caught without cash due to an emergency such as an ill family member who needed medicine?" A system that was flexible and transparent could also be fair and compassionate.

He described a system that a nearby community uses in which warning letters are sent out to those who have missed a payment. If the payments continue to be missed, a final warning letter is sent. Ultimately, turning off the water service was used as a last resort.

As we ended the meeting, I was reminded how important all the decisions are in sustaining a community water system when I remembered another community Don Mike and I had worked in. The community had turned in a solicitation for a new water system to Don Mike and when we visited them, they insisted that they had never had a water system and had always walked two kilometers to the river to fetch contaminated water. But something did not seem right. We kept seeing signs of an old water system and finally, when we visited the spring site, we uncovered an abandoned pipe for an old system.

Don Mike exploded and demanded that a full history be provided.

It turned out that a system had been built seven years prior, but had failed. The water committee looked the other way when their friends and family fell in arrears on their water payments and the system did not have sufficient funds to keep up with the repairs. People got frustrated and more refused to pay.

"Why should I pay when I only receive water three days a week?" One user had said.

The system fell into complete failure, forcing everyone to fetch water once again from the river.

Because of the dishonesty, Don Mike declined the solicitation. Fortunately for the community, they were located near Antigua and another non-profit picked up the project not knowing the history. Sadly, only three years later, Don Mike received another request for a water system from the same community as the system had failed once again.

Knowing how important it was to have informed consent on the important decisions that needed to be made, the committee next held a series of community meetings. At the first meeting, the people filed into the school, looking forward to an update on the project's schedule as they were all anxious to find out when water would finally arrive at their homes. Don Mike remembered that after the updates were given on the project's progress, Rolando explained that they had several important decisions that needed to be discussed and made.

"These decisions are not going to be easy," he said. "But they are critical to get the water system off to the right start if it is going to provide reliable water service for our children and grandchildren."

Rolando outlined the list of topics that needed to be discussed. One by one, the topics were debated, and Don Mike later informed the students of the decisions made.

- A plumber was appointed and would be paid.
- A monthly fee of Q40 ($5.50) would be paid and meters would not be installed at this time. This would equate to about 4% of an average families monthly income.
- A connection fee of Q750 ($100) would be charged. Future connections would be charged the same if the family participated in the construction work. Connections for new families would be charged Q8,000 ($1070).
- A treasurer was appointed to keep track of payments and expenses, which could be inspected by any user.
- The water committee chairman would be responsible for sending out warning letters and the committee would go as a group if a service needed to be turned off.

At the last meeting, one member of the community asked, "Why is a fee needed at all? God gives us water and it should be free."

Don Mike responded, "Yes, God gives us water for free. But it is in the river and contaminated with bacteria that causes sickness. God does not provide clean water at your home for free. If you want clean water to keep you and your family healthy, a water fee is needed to sustain the system."

Rolando thanked Don Mike for his coaching through the difficult decisions that needed to be made.

"I see now why it is so important to make these decisions before we begin construction," said Rolando. "I think that if we had waited until after construction, we would have never truly engaged the people in the discussion."

With these important decisions being made, the community was now ready to move into the construction phase.

Don Mike at Don Guillermo's hotel in Joyabaj

CHAPTER FOURTEEN
Building the Water System

"Every project has its unintended consequences."
Michael Shawcross

Katie was as committed to the project as anyone on the team – maybe even more so. Even though her two-year Peace Corps assignment was up, she applied for a one-year extension with the Water Committee of La Garrucha as her counterpart organization. She had fallen in love with the community and was committed to do whatever she could to help them, including putting her life on hold another year for the project. Her request was granted, and she moved to La Garrucha to being organizing the project's logistics.

A lovely new home that was not yet occupied by its owner was selected by the community near the center of the town for Katie. Mincho and two masons working on the construction of the project would also stay in the home. It was painted bright turquoise and had plenty of room for the tools, materials and team members. A perfect home and construction yard.

The community then approached Don Mike and asked if they might apply to a federal government program to pay for their water connection fees. The program was called FONAPAZ and had been established as part of the peace accords to help impacted communities reestablish their infrastructure. Katie and Don Mike both advised against it, believing it would diminish the community's investment and likely increase the request for additional taps. Don Mike also pointed out that the program had a terrible reputation for being slow and bureaucratic.

But I argued that the community be allowed to apply. I knew they were still strapped financially from the damage done by Hurricane Stan and wanted them to receive all the help they could get. It was a decision that I would soon regret.

Don Mike had taught me time and again, the importance of community engagement. But I let my heart get in the way and apparently, needed another "lesson" on the importance of engaging the community in all aspects of the project, including its finances.

The community turned in its application to FONAPAZ. They were hopeful that the atrocities committed in their community during The Violence would improve their chances. The good news soon arrived! They were approved and FONAPAZ would promise to provide several kilometers of water pipe for the project within a month.

With the news that the project was indeed moving forward, Amy was inspired by Katie and considered putting her own life on hold and spending the summer after her graduation in La Garrucha to assist the construction team. I encouraged her, as it was a tremendous opportunity for a young engineer to build the project she also designed, a chance many engineers never get to experience in their entire careers.

Engineers Without Borders USA had never done a long-term placement of a volunteer up until this point, but Cathy Leslie, the executive director, also saw the value. With Cathy's health and safety concerns addressed since Amy would be staying with a Peace Corps volunteer, she willingly gave her approval. Amy was the first of many long-term EWB volunteers / fellows who would be deployed to countries around the globe. The impact of these volunteers has been tremendous, and it all started with Amy.

When Amy arrived, Don Mike warned her that she would likely receive several marriage proposals from the young gentlemen in the community. Amy blushed and doubted that it would occur. Don Mike knew that a young, athletic blonde from the United States would certainly not go unnoticed. On her second day, she called Don Mike and confessed that she had already received her first marriage proposal from a young man who gave her a ride on his motorcycle.

"I thanked him and quickly walked into the house because I did not know what to say," Amy confessed.

Don Mike laughed and suggested, "You might want to talk about a boyfriend back home. You know, you could also wear a ring."

In fact, Katie and Amy were the talk of the town in the community. People would stroll by in front of the house to catch a glimpse of these two foreign women. Women in the Highlands typically do not work construction and lead engineering projects and the people were curious, especially the young girls.

Construction

The community was anxious to get started. It was the rainy season and the water trench digging would be easier in the wet soil. Of course, they also wanted to have the water flowing before the next dry season. A groundbreaking ceremony was arranged at the spring with all the appropriate dignitaries and speeches.

It was a perfectly clear morning when Rolando laid the first stone near the banana tree. After everyone sang Guatemalan National Anthem, they asked Amy to sing the American National Anthem. Amy put her years of choir practice to good use and belted out a stunning rendition which was met with a rousing applause.

With the first stone laid, Rolando was hard at work arranging the work crews. As the loads of materials were delivered to the construction team's house, the community's excitement grew. Stacks and stacks of pipe, cement and valves were piled up, ready for installation. With all the materials stacked in one location, it was impressive, and it also gave everyone the sense of how much work was really in front of them.

The project would require nearly twenty miles of trenching to be dug across the valley and within the community to provide the water service. The trench would be two feet deep and a foot wide – all dug by hand. A concrete structure, or spring box, would be built to protect the spring and keep it clean. Two large concrete storage tanks would also be needed along with many concrete valve chambers. Two bridges had to be built to carry the water line across the rivers.

A group of a dozen workers were needed each day to work with Mincho and the masons on the concrete tasks. Larger work teams were organized to do the trenching and the community chose to work once or twice a week with crews of more than one hundred people to dig the trenches for the pipeline by hand.

I watched in amazement as the long line of workers left the house where the materials were stored and carried the pipe out to the field. Each worker was assigned a fifty-foot section of trench to dig for the pipeline. The digging was difficult as the workers cut through tree roots and used shovels, picks and hoes to move rocks the size of motorcycles. After digging the trench, the masons would follow up, connecting the pipes. After carefully placing the pipe in the trench and conducting pressure tests to ensure no leaks existed, the workers would backfill the trench, making sure that no rocks were against the pipe that might damage it.

Mincho inspected the work and was always encouraging the workers, singing and cracking jokes to help make the difficult work go easier. Sometimes, he even took a pick from a tired worker to give him a chance to catch his breath. With such large work teams, a mile of pipeline a day was dug and placed. Standing on top of the hill where the distribution tank would be built, I could watch the hundred people working and placing the pipe as it slowly progressed across the six kilometer-wide valley. I was amazed at what could be accomplished when 100 men worked as one.

On days the pipeline crew was not working, the masons with a smaller group of community volunteers constructed the two bridges that used cable to suspend the water line over the river. Anchors were poured and towers built out of concrete and rebar. The workers carried the materials, including the 94-pound sacks of cement over a mile to each of the bridge sites. The paths were narrow and slippery and the workers carefully made their way along the streams to the bridge site as they dared not drop one of the precious sacks of cement. The value of a sack of cement was more than what a typical plantation worker might hope to be paid for a day's work.

After nearly a month, the pipeline reached the site of the distribution tank. The valve was opened near the spring and a small group of workers and families waited anxiously for the water to arrive. After a few minutes, a gurgling sound came from the pipe.

"It's coming, it's coming," Amy exclaimed.

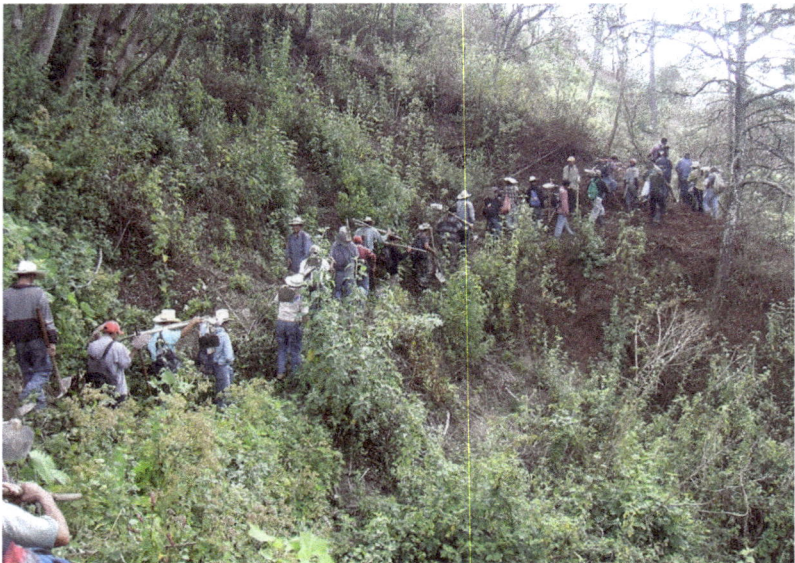

Workers head out to dig the water line trenches

Then the water started to flow out the end of the pipe and everyone cheered. The children jumped up and down, not fully understanding the significance of the event, but they could sense the excitement of their parents. The water was now in the community and a major milestone was achieved after two months of hard work. Women could now come to the end of the pipeline to fill their buckets and taste the water.

Now that the water conduction line had been completed, the workers focused on the distribution system that would provide the household water connections. The work would start at the top of the community and work its way down the mountain. First, the two large water tanks needed to be built that would be used for storage as the workers mixed and placed concrete day after day.

Next the sixteen kilometers of distribution trenches were dug for the three sectors of the system. The trenches wound their way through the community as they connected the water line to each of the 157 households. The work would focus on one sector and then move to the next. The excitement grew. The work teams were easier for Rolando to recruit as the work was right outside the community's homes and visible to all. The women cheered on the workers as they swung the heavy picks into the soil, offering refreshments of coffee and tortillas. The children enjoyed playing in what was likely the "world's longest sandbox" as more fresh dirt was generated each day.

As the first sector was completed, the homeowners asked if the water could be turned on to the homes. Don Mike, Mincho and Rolando insisted that everyone would receive water at the same time, not sector by sector. Don Mike had suggested that this be done to keep the workers focused on the community aspect of the project and keep the work teams large and motivated. The strategy worked as the work teams were even larger as they moved to the second sector.

The work progressed six days a week with Sundays reserved for laundry and rest.

"I never knew laundry was so much work," exclaimed Amy as she scrubbed her cloths and wiped the sweat from her brow. "This takes forever, and a lot of water."

In the evenings, Katie and Amy would join Mincho and the masons around a wheelbarrow filled with sand that served as a makeshift fireplace. As they all watched the fire burn, the team exchanged cultural information.

"Is it really true that it gets so cold that lakes freeze?" the masons asked.

Katie and Amy laughed.

"Yes, and the snow will get more than two meters deep at my house in Michigan," Amy explained to the wide-eyed listeners.

Rolando was curious if Americans played *fútbol* the same way that they did. Soccer was the sport of choice and everyone, young and old, participated in the activity. On Sundays, a pasture would be temporarily converted into a soccer field with everyone from toddlers to grandparents participating.

Amy tried to explain that many parts of America's northern region played a different sport called hockey. As they pressed her to explain, she wrinkled her nose trying to find an explanation that they would understand.

"It is a lot like *futbal*," she said, "but, you don't play with a soccer ball. You use a hard rubber disk instead called a puck."

Puzzled, the masons tried to imagine this strange sport.

She continued, "The disk is hit into the goal, not unlike *futbal*, but the goal is much smaller. The players also use wooden sticks instead of their feet to direct the disk." She stood and demonstrated hitting a block of wood with a board.

They all laughed.

"But the most interesting part is that the game is played on a field of ice and the players have to wear skates to move about the field." She sat back and watched the masons look at each other, shaking their heads and trying to visualize this strange sport called hockey.

Guatemalans love to catch and eat fish and they were very curious if there was fishing in America.

Katie explained that there was, and even in the winter, one could go fishing on the frozen lakes. She explained how fishermen would walk out on the frozen lake and chip holes through the ice. Fishing line would be dropped down through the hole to where the fish waited.

Much to the listeners' delight, she enacted the process of pulling up a fish through the ice-fishing hole with her arms flailing in the air as she demonstrated the process of pulling up the fishing line by hand. She then explained that on some lakes, the ice would form thick enough for people to drive their pickups out on the lake.

As the fire in the wheelbarrow burned low, the people howled and laughed, saying that they now knew she was telling a "tall tale" as ice could never hold up a truck.

As the work moved forward towards completion, the waterpipe to serve the homes had still not arrived from FONAPAZ. It was months late and the officials kept promising that it would arrive the next week, but it never did. Don Mike reminded me that this was a usual delay tactic. Rolando and the water committee made several visits to the FONAPAZ office, only to be told the same answer – next week for sure.

Finally, out of frustration, I decided to visit the office. I dressed up in my finest clothes and even put on a tie to try to impress the FONAPAZ officer. Rolando, Katie and I arrived at the office and were immediately rushed into the office instead of standing in line with the others waiting for service. I was impressed and was sure that the community would now get their pipe.

The FONAPAZ officer exchanged pleasantries with the group and sent for his boss to meet with them. Soon the Deputy Director arrived and gave each of us a Coca Cola. He then apologized that the pipe was delayed and promised that the order would be delivered next week.

I cleared my throat and stated this was the same message that we had been told for months, and still we had no waterpipe. I pressed the Deputy Director on which day the pipe would be shipped.

The Deputy Director's face tightened, and he said, "I think I have already told you that the pipe will come next week." He then pushed back from the table and left us alone in the room.

Later, after Don Mike heard the story, he told me, "I am afraid that you likely made things worse with that stunt."

It would be six more weeks before the pipe arrived to complete the home connections.

Finally, the day came when the final connection was made to the last house on the last sector. The workers looked at each other. Could it be true that all the work was completed?

It was time to turn on the water!

Rolando asked all the women to go to their homes and turn on the tap. He and the masons then turned on the water at the storage tanks located on top of the mountain at exactly the same time so the water would flow out into the different sectors and waited. The water flowed out of the full storage tanks and soon reached the first home. Standing at the storage tank, I could hear the woman squeal with delight. Then the water reached the next home and then the next. Cheers of joy rose up throughout the valley as home by home water reached them.

Don Mike stood by the tank and I saw a tear flow down his cheek and into his beard. He had opened many water systems before but said, "Every time is like the first time and the feeling never gets old."

Three girls and their parents celebrate the arrival of water

A Party is Planned

Gavina was overwhelmed with joy as the water flowed out of her tap. For as long as she could remember, she had to fetch water for her family. It was the same for her mother and her grandmother. She stared at the water in disbelief and then reached her hand under the tap and let the cool, clear water flow through her fingers. She started to giggle like a little girl.

She knew how the water would change the health of her community forever. The frustrations that she had felt for so many years as a midwife without access to clean water were evaporating. It was no easier for a midwife to do her job without water than a carpenter without wood, or a farmer without seed corn. She now felt that she was finally in a fair fight for her community's health and she was determined to win.

When Rolando entered the home, she raced to him and gave him such a hug that he could not breathe. He laughed and squeezed her back, knowing how important clean water was to her. As she continued to cling to him, she thanked him for all of his hard work and dedication. She was so proud of him for removing the community's curse – the lack of clean water. Their lives would never be the same.

Surely a party like none other must be held for such a monumental event. The community held meeting after meeting to plan the affair. It would consist of a three-day celebration and all the neighboring communities would be invited. Despite the considerable expense, 150 chickens would be served, and three young bulls would be butchered. On the first day, the Mayan priest would conduct the ceremony and give thanks. On the second, the Catholic priest would have the honors and on the third day, the dancing would be only interrupted long enough to get another glass of the local moonshine that filled a fifty-five-gallon drum.

The first day of the party was clear and crisp with a few gentle clouds in the sky. The people came from miles around to help La Garrucha celebrate, knowing what an important event it was for them. The men talked and complained about the crops and the price of corn. The women made piles and piles of tortillas as they talked and complained about their men. The children gathered in the school courtyard to play games that looked like total chaos, but were somehow understood by the participants. At the end of the day, some even slept at the school to not miss a minute of the next day's adventures. Finally, after the third day, they all retreated to their homes, exhausted.

Unintended Consequences

Don Poncho held conservative beliefs and protected his daughters fiercely. No young man in the community even dared look at them, let alone ask if he might talk with one of the dashing young ladies for fear of the wrath of Don Poncho.

During the groundbreaking celebration, Don Poncho and Lucinda's oldest daughter was introduced to a young man from a neighboring community by a friend. He was so wonderful and handsome that she couldn't say no when he asked if she would walk with him for a while. On the weekends she would join her girlfriends by the river, knowing that her suitor would be there waiting. They would walk along the trails by the river, talking quietly as the other girls giggled nearby.

One night, they decided to elope. When he found out the next day, Don Poncho was furious and hurt. Despite Lucinda's protests, he pledged that she would never be allowed to set foot in his home again. She was dead to him now.

Tatiana, Don Poncho's second oldest daughter, watched this all unfold. She had never seen her father so angry and hurt as he stomped around the house.

One day Tatiana noticed one of the masons working on the water system. He introduced himself as William and offered to carry her heavy jerry can filled with water that she was lugging up the mountain path to her home. He easily swung the jerry can onto his shoulder like it was as light as a feather and escorted her home. She was so nervous; she could not utter a word as he tried to chat with her. When they got within sight of the house, she quickly took the jerry can from William and ran into the home before her father could see her with him. She knew he would not be pleased.

We were frequent guests at Don Poncho's home during the construction of the project. We would discuss the project after dinner with Don Poncho and pass the evenings telling stories of days gone by. Tatiana watched William carefully as he was always the perfect gentleman. He helped clear the table and even offered to do the dishes from time to time.

One evening, William whispered to her that he would be by the river the next afternoon and asked if she might want to stop by and go for a walk. She quickly checked to make sure that Don Poncho was not watching and was able to somehow respond with a "yes" as her heart nearly stopped.

The next day she impatiently waited for her girlfriends so they could go to the river without drawing any attention. William was waiting for her and asked if he might hold her hand as they strolled up the river path. As his strong but gentle hand cradled hers, she knew that they were soul mates.

Tatiana and William could hardly wait for Sundays when they could meet again by the river and talk about their hopes and dreams. Don Poncho had not noticed any change, but Lucinda knew that something had changed with her daughter as Tatiana skipped around the house.

Tatiana and William made plans to elope as soon as the water project was completed. Tatiana knew it would devastate her father but feared that he might somehow stop her from being with William if they asked his permission to marry. She knew he was conservative, and it was a risk that she was not willing to take.

The night after the community's celebration, William and Tatiana eloped, going to live with William's family.

Don Poncho was beside himself. All the joy and happiness that he felt the day before was gone. He was crushed as he felt that he had lost a second daughter.

As the weeks turned into months, Tatiana repeatedly sent word home asking if she and William could stop by. Lucinda's reply was always that her father needed more time. Maybe in a few more weeks.

Nearly a year later, William was working on a pedestrian bridge over the Rio Motagua with Mincho and me. As we finished the project, William asked if he might have the left-over cable that was not needed for the project. I thought it was a strange request, but agreed. William gathered his friends and they carried the left-over cable up the river several miles to Don Poncho's home. When I heard the story, I laughed. "That's the first time I have ever heard of using cable as a peace offering."

A year later, I was invited to William and Tatiana's home for dinner. William had just completed the construction of a primary school that I helped Katie design and build. William proudly showed off the new school along with the construction of the small little block home he had just finished for Tatiana and him to live in.

As I entered the house, I noticed a first grade reading book sitting on the table. I joked with William that maybe he had gotten the reader for a new baby that was on the way. But then realized that the reader was for Tatiana.

Don Poncho had not allowed Tatiana to go to school, insisting that she help her mother by fetching water and gathering firewood.

Tatiana had longed to go to school her whole life, but willingly complied with her father's direction. Now that the new primary school in their community was complete, William encouraged her to start going to classes.

He knew how important it was to her and encouraged her by saying, "It's never too late to learn."

So, on the first day of class for the new school year, Tatiana proudly walked into the school and sat down with the other first graders to learn how to read.

As William told the story to me, I imagined Tatiana sitting at her small desk with the other first graders learning to read. I could only imagine how brave she must be. I was so proud of them both and happy that the little school was having such a big impact on her life. I was also so thankful that in La Garrucha, no more girls like Tatiana would be denied the opportunity to attend school because they had to fetch water.

Don Mike attends another community meeting
Photo credit Scott Mitchell

CHAPTER FIFTEEN
Water Utility Post-Construction Coaching

"Water utility projects need post-construction coaching to be sustained.
There is tremendous value in offering unbiased, third party support."
Michael Shawcross

Don Mike had been helping communities solve their drinking water needs for more than forty years. He knew that once the system was built, his job as a coach was not over and the challenges the community would face to sustain the system were just beginning. He knew and accepted that this required a long-term commitment on his part to the community.

In La Garrucha, it was the same. Don Mike talked with Rolando and the water committee and let them know that he was always more than willing to help them through the rough spots that were sure to come. It was their project and he expected them to operate and maintain it, but he would "walk with them on their journey."

I observed that Don Mike, like any good coach, adjusted his style to the members of his team and the situation. Sometimes he needed to offer some encouragement when the committee was down. Other times he needed to provide some advice, and sometimes a swift kick in the butt when the committee knew what to do but was too scared to do it. Most of the time, he simply needed to provide moral support and encouragement.

It seemed that Rolando and the water committee knew what to do and how to do it, but from time to time they needed someone to help convince the community that the decisions were reasonable. Once or twice a year, Rolando would call and ask if Don Mike could attend a community meeting that might need his support and advice. The Water Committee held community meetings once a quarter or if a crisis demanded immediate attention.

The meetings always started with a report out on the water systems financial status and any issues that had been brought forward to the water committee by the users. Many times, a user or group of families within the community would voice their dissatisfaction with the fees and how they were being collected.

Typical questions that I had heard the community members ask included:

Why are the fees so high?

Why do you publish who has not paid the fee and embarrass us in front of our friends?

Why did it take so long to repair the broken line?

Why do you have to turn the water off for so long when you clean the tank?

Being a member of the water committee was thankless work that had no compensation and generally few friends. It really was community service and sometimes I wondered why anyone would ever volunteer to be on the committee.

Don Mike viewed that his job was to support the water committee and encourage the community to work through its issues to an acceptable resolution. Don Mike was respected by the community as he had worked with them to accomplish two major projects that had been a barrier for generations to the community's goals.

The power of his unbiased, third party advice could not be overestimated. He had nothing to gain from the systems operations and maintenance. He was not going to get paid, run for political office or receive a promotion at his business. It was clear to the community that he was there only to help.

The community meetings were typically well attended. Don Mike always wondered if it was because people really cared that much about the water utility or if it was the best form of entertainment available that evening.

At a typical community meeting, the committee would allow the complaining party to voice their issues and concerns. This could go on for some time as the points were repeated over and over again, but the speech was never interrupted. Then the water committee would give their explanation of the situation and might offer up adjustments to the utility that could be voted on by the community.

Many times, someone in the audience would ask Don Mike what he thought. He would always remind the attendees that he did not live in the community and did not have a vote in the matter. He would thank them for inviting him to the meeting and was always happy to visit their beautiful community and catch up with so many of his friends. He would then always find something positive to comment on that he had seen.

"I love the new vegetable garden at the school." Or, "It looks like you have been working hard on the roadway's drainage. It looks great."

He always reminded them that he had no vested interest in the outcome. On most occasions, he then explained to the community that the water committee's action was typical, reasonable and fair in his eyes. But they had no obligation to agree with him, as it was their decision and he was there as a guest. He would always end his words with encouragement and hope.

"I am so glad that you all feel so passionately about your water system. It is a blessing for everyone to have clean water and I know it is a struggle to sustain it. But I know you all want the same thing – a functioning water system that will provide its many benefits to you and your children and their children in the future."

Generally, things were going well until the election for the new mayor was held. A candidate was running against the incumbent on the platform of potable water. Don Mike was so excited that potable water was now elevated to such an important political issue, but he never dreamed that it could jeopardize the sustainability of the nearly 100 water systems in the municipality.

Then Don Mike learned that the candidate was not only promising to make potable water a priority for every community, he was also committing to pass a law that would make it illegal to turn off anyone's potable water supply. Don Mike's heart stopped knowing the seriousness of the situation.

He quickly scheduled a meeting with the candidate to explain the risks of such a promise. As Don Mike and the candidate met, Don Mike urged him to back away from the promise with his passionate plea that "clean water is not free."

Why would anyone ever pay their water bill if there was no consequence?

How would any of the systems ever have the financial resources needed to keep up with the maintenance and operations?

Could the municipality pay for all the operations and maintenance costs for the 100 rural systems?

Don Mike then described the "water death spiral" that is caused by the cycle of poor service that results in fewer people willingly paying the fees, which in turn results in even poorer service. The cycle repeats until the system completely collapses.

The candidate listened intently and said. "Thank you Don Mike for the information and sharing your years of experience with water systems with me. You have given me plenty of things to think about and I now understand that operating and maintaining water systems is not as simple as I had thought."

One day, Don Mike and I were in the square as the election drew near. There was a large crowd gathered to hear the candidates and it was sure to be a close vote. As we watched the new candidate stump in front of the crowd, we couldn't believe what we heard.

Clearly, the candidate had chosen to ignore the advice of Don

Mike and make political promises that would be popular with the people. The crowd cheered as the candidate promised to pass a law to throw anyone in jail that turned off a resident's water service.

How could anyone be opposed to free clean water?

Then the incumbent mayor took the stage. He knew that the municipality did not have the resources to pay for the operations and maintenance of the 100 rural systems. He had gotten elected based upon his honesty, and he would not lie now. When he was pressed on the topic of free, clean water with a question from the audience, he tried to explain the importance of a fiscally sound water utility to make it sustainable for all. The crowd reacted rudely, and we knew it was over. The next week, the candidate won the election and became the new mayor.

After the election, the new mayor followed through with his promise and passed the law making it illegal to turn off a home's water service. Rolando and the water committee chose to continue to follow their management practices of collecting the fees and sending out warning letters. A few months later a homeowner challenged the water committee's authority to turn off his water supply and refused to pay the bill. The homeowner had heard of the new law and decided to test the water committee.

Rolando remembers the next water committee meeting, knowing that this was an important challenge to the water utility's operations. They struggled with what to do and decided to respond as a group. The next morning, the entire water committee trudged up the mountain to the man's house. They knocked on his door and gave him one more chance to reconsider and pay his water bill.

When he refused, the water committee gathered around the valve box and promptly turned off the service. By this time, a small crowd had gathered to witness the outcome. They clapped as the service was turned off and the homeowner stormed back into his home.

It seemed that the crowd wanted to make sure that everyone in the community was being treated the same and no exceptions were being made for "a complainer."

The next day, a black pickup with the yellow words "Municipal Police" painted on its side drove up to Rolando's home. The rest of the water committee was summoned, and they explained to the policeman the need to collect fees for the water system. The policeman nodded his head but explained that his job was to enforce the law, and in this case, the law was clear. He then promptly arrested the entire water committee! All of them!

As the committee sat in the back of the pickup, departing for jail, Rolando instructed his wife to contact Don Mike.

When Don Mike heard the news, he was furious. He immediately jumped into the Landcruiser and sped to the municipal jail, where he demanded the immediate release of the water committee. After some negotiations, he agreed to pay a bond and the water committee could go home.

At the court date, Don Mike attended the proceeding to lend his support. The judge was sympathetic and understood the logic but was also in a position where he could not re-write the law. He allowed the water committee to go home with "time served" and a promise that they would not turn off anyone else's water service. They were happy to return to their families but knew that they had a major challenge in front of them.

As the water committee met with Don Mike and me back in Don Poncho's church, they pondered their predicament. They were frustrated, angry and embarrassed. It was clear that they had to follow the new law until they could get it changed, but that might take years. In the meantime, they had to somehow restructure the utility so that it could be sustained. They had come too far to let the system fall into the "death spiral" that Don Mike so often referenced.

They felt that they had to lower the monthly fee as much as possible to minimize the collection issue. To do this they decided to make the plumber a volunteer position that would no longer receive a salary. The position would be rotated every other year between water committee members to not burden only one person with the duty. The fees would be reduced accordingly.

This would require tremendous dedication and service by the volunteer water committee / plumbers, but they all pledged to make this personal commitment and sacrifice until they could get the law changed.

Next, they decided to comply with the law and not even threaten to turn off anyone's service. They would have to use peer pressure to push people to comply with paying the fee. If someone refused to pay, the water committee would need to recruit friends and family of the delinquent user to help convince them that it was their civic duty to pay their fair share. It was the best they could do under the circumstances.

At the community meeting to approve the water utility changes, the complaining homeowner was not in attendance. Many community members expressed their apologies for the behavior of their neighbor and that the committee had to go to jail. They understood that in light of the new law, changes were needed, and they would support the water committee in its efforts moving forward.

As Don Mike left the meeting, he was concerned that clean potable water could be sustained by the volunteer plumbers. But it was not his decision. It was their water system and they had reached a consensus on a path forward. For this reason alone, he was proud.

I also saw how all the hard work of coaching the community had paid off. The community was able to recognize and address a very difficult issue. Without the acceptance in ownership and the foundation of trust in the water committee, I knew the system would have likely failed as had so many I had seen before and since.

Hurricane Agatha

In June of 2010, at the very start of the hurricane season, Hurricane Agatha ravaged the Highlands of Guatemala. The storm stalled over the Rio Motagua valley causing extensive flooding and landslides. Some local areas reported that more than forty inches of rain fell from the storm.

Just a few kilometers upstream of the bridge at La Garrucha, an entire side of the mountain slid down into the valley and created an earth dam across the Rio Motagua. Soon the river filled the area behind the landslide with water and overtopped the earth dam. The loose material was no match for the rushing water and the entire pool of water was released down the river valley. It was a dam break with a giant wave of water, mud and debris rushing down the valley.

It appeared that every bridge was destroyed by the wall of water from the dam site to the ocean, a distance of more than 500 kilometers, cutting the eastern portion of the country nearly in two. The damage was extensive and Florencio, who was now the mayor of Joyabaj, called on Don Mike and me to help with emergency access and water system repairs similar to Hurricane Stan.

Florencio knew his people were hurting and had seen the work Don Mike and I had done five years earlier during Hurricane Stan. Of course, Don Mike had already begun preparing the Landcruiser even before the last raindrops fell. He promised Florencio that he would do all he could to help.

I quickly contacted two churches, mine at Peace Lutheran and Paul's at Southminster Presbyterian in Waukesha, Wisconsin, to raise some emergency funds. Several Rotarians also chipped in on short notice to help raise a pool of disaster relief funds for water and bridges. I quickly transferred my work tasks to my willing peers and kissed my wife who said, "You and Don Mike take care of each other."

By this time, she had worked on several projects in Guatemala with me and Don Mike. She had grown to love the quirky old Brit who drank and swore too much. "It is no wonder he is still a bachelor," she said as she had looked at his house for the first time and shook her head. She also knew that my mentor held a special part in my heart and this work was important to me. She would support me now, as she always had.

At the airport in Guatemala City, a newspaper caught my eye as I scurried about with my luggage of tools. There was a large photo of Rolando proudly standing in front of the community's bridge across the Rio Motagua with the caption "The Unbeatable Bridge." I dug out a few Quetzals and purchased a copy.

The story explained that it was the only bridge to survive the rushing water of the Rio Motagua caused by Hurricane Agatha. Rolando was then quoted as saying, "the reason the bridge stood was because it was built the way it should be."

As Don Mike and I met with Florencio, Florencio explained that the municipality had more resources available this time than after hurricane Stan.

"I have resources. I just don't have enough resources to make any mistakes," he explained. We were impressed and promised to do our best to help with the response.

Photo in Nuestro Diario newspaper showing the Rio Motagua Bridge.

The first task was to restore an emergency crossing over the Rio Motagua where the major highway bridge had existed and connected Guatemala City to the municipality of Joyabaj. As luck would have it, a hospital convention had been held in Guatemala City and many of the doctors and nurses were stranded on the south side of the river, unable to reach their patients at the Joyabaj regional hospital.

A cable zip line was established to move one person at a time across the river and we proceeded to move the doctors and nurses across the river.

"Just keep your eyes closed," Mincho yelled to one nurse who trembled in the rope chair that held her above the raging river as he pulled her across.

After the doctors and nurses were safely across, we carefully pulled boxes and boxes of medical supplies across the river using the same zip line assembly.

Nurse crossing the Rio Motagua using a temporary zip line.

Similar to the work post Hurricane Stan, we initially focused on bridges so that emergency supplies could reach the areas most impacted. We applied our previous experience to the new disaster and quickly opened roads by constructing temporary repairs or building temporary bridges to allow the heavy trucks to reach those in need.

Next, we turned our focus to water. It was June and thunderstorms occurred nearly every afternoon. Although the rains made moving materials and equipment difficult, it also allowed the people to harvest water from their roofs.

This was fortunate, as nearly every water system had sustained some damage and the dry season would come soon enough. Applying the lessons learned from the Hurricane Stan work, we systematically worked through the repairs, walking the lines, developing the designs and ordering materials for the repairs in a process that took only a matter of days from start to finish at each community.

We worked hard to stretch every dollar, but finally the funds ran out.

Florencio had received word that a large water system that served the 20,000 people just north of La Garrucha needed to be repaired. Don Mike and I knew the community well, as it had sent many workers to help build the bridge over the Rio Motagua five years earlier.

When we arrived, we could see that the community was surviving by harvesting rainwater from their roofs, but that would no longer be an option when the dry season arrived in a few months. The waterline repair would require a new 500-foot-long, high pressure water line bridge over the Rio Motagua that would cost $10,000. As we explained the repairs to the community and the fact that we were out of funds, the faces looked distraught.

Ten-thousand dollars seemed like an impossible figure to raise for them and the possibility of having to suffer through the dry season without the water service scared them. Then Javier, the village president spoke. He acknowledged the large sum but reminded the community that they were 20,000 people strong.

"We have each other and God has given us life. If each of us only contribute fifty cents, the sum will be raised. I know we can do it!" he exclaimed.

After some further discussion, the community agreed to raise the funds for the project.

As Don Mike and I left the meeting, I shook my head. These

people had already been through so much and many had already lost everything - including their home and even family members. It seemed impossible for them to raise the funds for the water bridge.

I had asked my friends and family back home to give all they could and would not go back and ask them for more. Cathy and I had stretched our own finances, using our donation to match the donations of others. Now, I was trying to find the words to explain to Cathy that we may need to take out a second mortgage against our home to fund the project and wait for the community to pay us back.

"How else would these people get the clean water they so desperately needed?" I asked Don Mike.

Don Mike looked at me and winked. "you just wait."

Ten days later, Javier called and said they had raised the funds. "We are ready to get to work. When can the materials be delivered?" Javier had walked to each and every home in the community asking them to contribute what they could towards the project.

When people pushed back, he said, "Maybe go without your cell phone for a month and use the money to fund the water bridge. Isn't water more important than your cell phone?"

Apparently, nobody could resist his persistent sales approach.

A Visit to La Garrucha

Now with the water bridge work well underway, Don Mike and I finally had time to visit our friends at La Garrucha just a few kilometers away. We knew that the area had been hard hit by the storm and there were reports that many homes had been destroyed and several lives had been lost. But we had not been contacted by Rolando or anyone from the community asking for help, so we did not know what to expect.

We approached from the north, making our way down the

roadway in the trusty Landcruiser that we had driven so many times before. The roadway was hardly recognizable with all the landslides and damage. We made our way down from the top of the mountain into the Rio Motagua valley, dodging large boulders that had fallen onto the road. As we rounded the bend, we saw the bridge standing, just as in the photo in the newspaper. When we reached the bridge, the homeowners nearby came out to greet us and tell the story of the wave of water that rushed down the valley after the dam broke. They had seen the river's flow go down and knew something strange was happening upstream.

Then, they heard the roar of the river as the huge wave raced down the valley. They had no place to go and it was raining hard, so they huddled in their homes and prayed. The water had reached the foundation of a few of the homes, but everyone was safe. As they looked out their doors, they could see the wave covering the bridge with five feet of water. Trees tumbled down the river like matchsticks and passed over the bridge. Surely, they felt, the bridge must have been destroyed, but somehow it withstood the force of the river and stood strong.

For years, I would be contacted by agencies and organizations asking what the secret was that allowed the bridge to stand. The legend grew with stories that the bridge had been blessed with a special Mayan ceremony. Another legend was that the concrete had been mixed with bull's blood to make it stronger and that is how it survived. In the end, I always admitted that it was dumb luck, nothing more.

After inspecting the bridge, we next moved up the mountain to the center of La Garrucha to see how the water system had fared. We saw landslides and devastation along the way, and we could see that the roadway had been cleared using hand tools. There were several places where we could see that the water line had been damaged, but in each instance, the water line was already repaired.

One of the water line bridges over the Cujil River had been destroyed, but a temporary line was already in place. As we stopped by the school, which was back in session with the teachers and most of the students present, Don Mike turned on the tap at the handwashing station and the water flowed. Don Mike did a little jig, much to the student's delight.

I was stunned that the water system had been restored already.

The community had recognized the Landcruiser and word spread that Don Mike had come to visit. As we met with our old friends in Don Poncho's church, they told the stories of the terrible storm and the damage that it had caused. Don Mike praised them for their leadership and ability to do the water line repairs on their own.

I saw Rolando look at Don Mike, surprised and maybe a little hurt. "There is no need for you to thank us. You told us that you would help with building the system, but we needed to maintain and take care of it on our own. We only did as we promised, nothing more," he said.

Don Mike beamed, now even more proud of this amazing community.

Once again, I marveled at the lessons of Don Mike and how community development done right, can instill ownership and sustainable results.

Don Mike cutting the ribbon to another new school at Mirrador

CHAPTER SIXTEEN
Encouraging Community Leadership

"Sometimes you only need to light the spark and get out of the way."
Michael Shawcross

After Hurricane Agatha, Don Mike's involvement with the community diminished. It wasn't that he did not care, or challenges did not present themselves. It was because the community had learned to overcome their challenges without outside intervention. Years later, I listened with great satisfaction to how the community tackled challenges and pushed itself to improve its own life.

Water System Improvements

Soon after Hurricane Agatha, Rolando knew he had to strengthen the water utility structure. The memory of him and the rest of the water committee being arrested for turning off a home's service was still fresh and there was little hope of changing the law soon. The volunteer plumbers were doing their best, but he was worried. They couldn't give up as they had come so far and accomplished so much. They must find another way to sustain the water utility.

For a few years they had been using a volunteer plumber to do the system's maintenance. This had reduced the monthly water fee on the users, making it easier for them to pay on a regular basis, but it was obvious that the work was too much to ask of a volunteer. It was clear that with time, it would be impossible to recruit the new plumbers needed every few years as they rotated the volunteer duties.

After some discussion with the water committee, it was decided to have more plumbers assigned to the system to lighten each plumber's load and reflect the volunteer nature of the work. Four separate plumbers would be used, and each would be assigned a geographic area of the system. The plumbers would also live in the geographic area assigned to improve accountability and minimize the time needed to check on the system.

They decided to continue the rotation of the position every two years with a six month overlap to improve the knowledge transfer of the system. Each plumber needed to know where the pipelines ran, where each connection was made for a home's service and what pipe flows each valve controlled.

The EWB Marquette students had provided a maintenance manual, but many of the plumbers did now know how to read so the learning would need to occur primarily in the field. The committee hoped that one advantage of the plan was that it would provide some redundancy, as the "retired plumbers" within the sector would also be knowledgeable of the system and could step in and help with repairs if needed.

The financial management of the system would be broken down into two utilities. The homes near the river had recently been given separate political status and were now their own *aldea* entity called "Motagua." They would not only have their own plumber, but also their own record keeping for who paid their fees and an accounting of the expenses paid. The other three plumbers in the Centro and Cujil sectors would work under a second financial system.

Rolando's biggest concern was the collection of the utility fees, now that they could not turn off any home's service. As time passed, it seemed more and more families were becoming delinquent in paying the fee. Eventually, they all would pay up, but it was taking more and more time for them to pay their bill. He knew he needed more support from the community if the system was going be sustained.

The committee decided to add representatives from the community to each water sector to pair up with the plumbers. This would provide a second person within each sector for homeowners to talk to, lodge complaints and gain understanding on the system's operations. The community representative would also provide an independent check on the work of the plumber in the sector and report any issues back to the water committee. The fee collection would be done by the community representative with the intent that they could also solicit help from a delinquent homeowner's friends and family if needed. The hope was by using peer pressure, the system could be sustained financially.

It was not what the water committee had hoped for when they set up the utility, but given the circumstances, they believed it gave them the best path to sustainability.

A Full-time Nurse Comes to La Garrucha

The community had waited a long time for a health care professional to staff the clinic full time. They understood the Ministry of Health's policy that staffing a clinic was not a good investment if the clinic did not have clean water. But now they had their water system and with it came the full-time nurse that would help them improve the community's health.

His name is Efrain and he had always known his calling to work in health care since he was a little boy. He had studied to be a registered nurse and was well trained in the health care system. His personality is an interesting mix of an intense health care professional with a large dose of compassion and concern for his patients. He demands a lot from himself and expects the members of the community to remain committed to improving their health. His dedication and compassion towards the community members soon endeared him to them.

Rolando related a story to me that demonstrated the community's willingness and ability to overcome difficult challenges. One day, soon after Efrain's arrival, news reached the community that due to a funding shortfall, the clinics across the country would be shut down indefinitely. As the community discussed this with Efrain, he explained that there was nothing he could do as the funding shortfall at the Ministry of Health would affect the entire country and was not targeted specifically towards La Garrucha. The community also appealed to the mayor, who explained that the country's clinics were financially supported by the Ministry of Health, not the municipality. He knew the impact on the people would be tremendous so he promised to lend his voice to theirs to see if they could somehow get funding freed up for this important service.

As Efrain was packing up his things and preparing to close the clinic, the community gathered outside the clinic and approached him. They asked if he would stay and continue his role despite the lack of funding for his salary. They did not have much, but would support him with chickens, beans and tortillas in hopes that he would stay until the funding was cleared up. Of course, Efrain accepted and continued to work at the clinic even though it was officially closed for several months.

Although it was a difficult time with limited medical supplies and plenty of uncertainty, it also made the community invested in the clinic and its operation. The event resulted in a close relationship that empowered and inspired them to do more, even after the funding was restored.

Building a Healthcare Team

As soon as Efrain arrived, he knew he was blessed with a team of knowledgeable and dedicated midwives who were already working within the community. They were led by Gavina who had been a midwife even before Efrain was born. She was one of the first midwives to gain official certification back in 1980 and continued to be the rock of the community's health care system as well as its biggest advocate. He realized that the midwives not only attended to expecting mothers and infants, but also were the community's front line health care workers responding to everything from a machete cut to a high fever. Everyone knew their midwife and trusted her to provide their primary medical care.

Efrain knew that integrating into the midwife's health care system was important. One of his first acts when he arrived was to meet with the midwives and respectfully asked if he might join their already outstanding medical care team. This simple act of respect went a long way to bringing the team together as one working unit towards a common goal of a healthy community. They were peers.

The midwives and Efrain meet every few days to review their cases and discuss the best course of action. I observed one of these meetings. It was a lively discussion of health care professionals, each weighing in with a possible diagnosis and suggestions for treatment. They learn from each other with each professional having their area of expertise, forming a robust medical team.

Now that the community had a water system which allowed for a full-time medical professional, the clinic also had additional medical supplies, including vaccinations, and equipment to help the medical team do their jobs. The midwives were interested in learning what procedures could be done at the clinic and which needed to be referred to the hospital.

The community medical team worked out a system where they did training twice a month. Once a month, Efrain would provide western training on a medical topic that the midwives had requested. Also, once a month the midwives would provide traditional medicine training to Efrain who was fascinated at how effective many of the local remedies were.

The system was so successful, that the classes were then opened to all the medical teams within the municipality and each meeting typically has over sixty attendees. Gavina beamed with pride as she told me about the latest workshop she taught using herbs to calm an upset stomach.

Advocating for the Midwives

Before the bridge was built, crossing the Rio Motagua was not an option for Gavina and the other midwives. Gavina would know when a mother was in trouble based upon her training and years of experience. She could feel the position of the baby and knew when things were wrong. Sometimes, she was able to turn the baby so a proper delivery was possible, but other times the baby would simply not be moved.

It was those circumstances that haunted Gavina. She knew she needed to get the mother to the hospital, but without access across the river it was not possible. She would not sleep during the days leading up to the difficult delivery, knowing that the mother was going to be in a fight for her life.

It was a fight that she knew all too well. She, herself, had a history of difficult pregnancies. Ten times she had been pregnant and had only four healthy children now in her family.

"Most times, a mother knows when things are 'not right,'" she said and she knew that feeling. She knew the anxiety of knowing that the upcoming delivery was going to be a struggle of life or death for her and her baby.

As the contractions would start for a difficult delivery, Gavina would try to be as optimistic as she could for the mother and the rest of the nervous family. She would send for another midwife to help her, knowing the difficulties that lay ahead. As the mother struggled, she would swear under her breath and through her smile. She was frustrated by only having limited ways of providing assistance.

As the delivery wore on, the family would become increasingly worried and would typically huddle in prayer as they would hear their mother's screams in pain. Sometimes she was able to save the baby and mother. Other times, despite her best efforts, she would lose the baby or the mother – or worse yet, both.

After a failed delivery, Gavina told me she would cry as she walked up the hill to her home. This was not just a patient. She knew these women and their families. It was a neighbor and a friend. She also knew the difficulties that lay ahead for the rest of the family when they lost their mother. Every child deserves a mother growing up and her sorrow would turn to anger as she knew that if she had access to a hospital, things would have been different.

Once the bridge over the Rio Motagua and the roadway to Joyabaj was completed, the midwives were very excited to have access to the regional hospital located there. The hospital had several doctors and a delivery room with the equipment that the midwives could only dream of. They knew that by simply getting their patient to the hospital the baby and mother's chances of success were vastly improved. It was a chance that they could now give their friends and neighbors.

Shortly after the bridge and roadway were open, Gavina was tending to a problem pregnancy that was coming to term. Her friend was her patient and had lost a baby previously. Try as she might, Gavina had not been able to turn the baby and she knew that they needed to go to the hospital. They loaded her friend into a pickup truck and she did her best to comfort her on the bumpy drive to Joyabaj.

As they arrived, they were met at the emergency room door and the on-call physician said, "Thanks, we will take it from here."

Gavina was forced to drop the hand of her patient and friend and was then escorted to the waiting room. She saw her patient pass through the door with a look of confusion and terror on her face. The mother only spoke Mayan Quiche and was in a hospital run in Spanish. She had never been to a hospital before and was terrified.

Gavina was upset. She knew the valuable history of the patient's prior deliveries, information that any doctor should know before attempting to provide assistance. She also knew how alone and scared her friend must feel as she looked at the sterile white walls of the hospital.

The next day, Gavina returned with the smiling mother and a healthy baby girl. When Efrain heard the news that Gavina had not been allowed into the delivery room, he was so upset that he traveled to the hospital and demanded a meeting with the Director. He explained that the midwives were an important part of the community's health care team and were key to convincing people to come to the hospital. They had known most of the expecting mothers all of their lives and had delivered many of them from their mothers. His passion could not be denied, and the hospital director agreed to allow the midwives to accompany their patients into the delivery room and assist with the delivery.

When he shared the news with the midwives, they jumped up and gave him a huge group hug. They had been so afraid that the access that they had been praying for would only prove to be of limited value if the mothers were afraid to go to the hospital. They now could try to convince every mother to have her delivery in the hospital with them by their side.

Adding Education to the Health Care Team

Now that clean water was available, I smiled as I watched Abraham, a teacher at the primary school, teach the children how to wash their hands. They proudly sang the "happy birthday song" together while they washed their hands to make sure they were washed long enough to get them clean. Abraham's classes had new energy and meaning now that clean water was available.

He knew the importance of hygiene, especially hand washing, and the benefits it would have for his students. Previously, he had dreaded the hand washing lessons as he was forced to teach the class by having them imagine they had access to clean water. Now, the clean water was real, and he knew the benefits would also be real to the lives of his young students.

He had almost immediately seen an increase in attendance after the water system was opened. He knew there would be benefits, but somehow thought that it would take some time to see an improvement in the students' health and attendance.

Soon after the water was turned on, new students, most of them girls, came to school for the first time. Since they no longer needed to spend the entire day fetching water from the river and firewood to boil it, their families could now allow them to go to school. The new students were of various ages, but all of them needed to be added to Abraham's first-grade class. The teachers gladly rearranged the classrooms to accommodate the new, eager students. Abraham was so excited to teach these new young ladies, knowing that an exciting new world was now possible for them.

Within weeks of the water system being turned on, he also noticed the improved health of his students. Students were no longer missing classes due to diarrhea and other water borne diseases.

Previously, nearly every student would miss several classes a month due to diarrhea. Even when they were in class, many were sick and could not focus on their studies. As their teacher, it was difficult to educate such young minds when their studies were interrupted, and the students could not focus.

Now, he looked at the room full of students and could feel their energy. He explained to me with excitement that this is why he had become a teacher and he could not wait to get to work.

Inspired, Abraham and the medical team consisting of Efrain and the midwives, developed an education plan for the community. The first step was to beef up the health classes in the school to include not only hand washing, but other health and hygiene activities such as food preparation, the cleaning of water containers and the importance of a good latrine. The students loved to color the pictures provided by Efrain that depicted good health and hygiene behavior and proudly took them home to show their parents.

The strategy worked as mothers soon came to Abraham asking if he might be willing to teach health and hygiene classes for them so they could learn like their children. He started offering adult health and hygiene classes in the evenings. Initially, they were attended by only a few mothers. But with time, soon the classes grew in popularity. The classes not only provided some important education, it also gave the mothers a reason to get out of the home and socialize with their friends. As I looked at Efrain and his new students who were laughing and carrying on, I could not help but be filled with joy.

Example of pictures provided by the Ministry of Health

A Community Health Committee

Gavina remembered that even though they now had access to the hospital, many members of the community still refused to go and seek medical attention. This applied especially to the expecting mothers, as many had never left the community and were terrified of leaving the safety of their home and making the journey to the hospital that seemed so far away. The fact that many of the adult women did not speak Spanish was also a major barrier.

As the Medical Team brainstormed ideas, the concept of a community health committee was born. The committee would be made up from the Medical Team, Abraham, and members from each sector of the community and report to the community's leadership. The community members of the committee would act as health advocates for their neighbors. They would also be an extra set of eyes to help the midwives identify those most vulnerable who may not be comfortable seeking out medical attention.

With the assistance of friend and family, not unlike the water committee members, the community health care committee members would work to convince and alleviate the fear of those who were scared to travel to the hospital. In many cases, they would accompany the person to the hospital and even arrange for the transportation free of charge. The idea worked and the number of children born in the hospital steadily grew year after year.

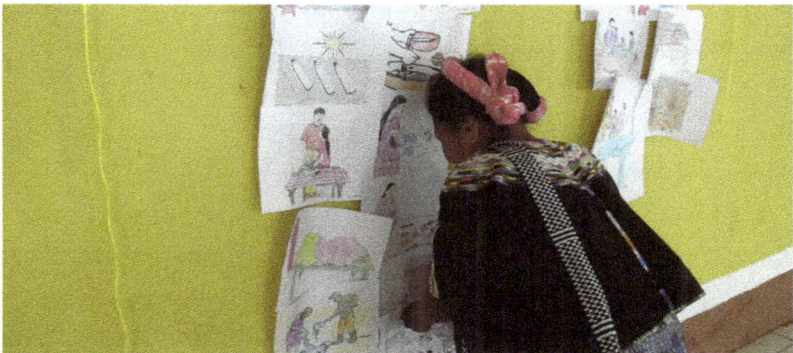

Adult health and hygiene training session

A Health Care Comparison

As the midwives explained their system to me, they asked me how medical services were provided in the United States.

"Was it similar to the hospital where the doctors only treated the body?" they asked.

The idea of only treating the body seemed so foreign to them. "Didn't the western world realize that a person is composed of their body, mind and heart and that all three must be treated to achieve good health?" they asked.

The midwives explained that most of their work was done by making house calls.

"Yes, the home visits took time," they explained, "but it is important to treat the mind and the heart along with the body." This allowed them to talk with the patient in the comfort of their own home, especially the women, and identify other issues of concern. They could also observe the condition and activity of the home and probe their friends and neighbors about other problems that might be occurring.

They shook their heads, as they said they did not understand western doctors' methods of only treating the body after a short consultation with the patient. "We are all people made up of our body, soul and mind. Each part is connected to the other and it is impossible to treat only one part," they explained.

I wondered if modern medicine could learn a thing or two from these dedicated ladies.

As I reflected on the initiative that the community had taken to sustain its water system and improve its health, I was reminded of the phrase that many communities would use, saying a project was "breaking the barrier to their development dreams." Clearly, La

Garrucha willingly assumed responsibility for its development after the barriers of lack of access across the river and to clean water were removed. They knew what to do to achieve their development goals and possessed the will to do it.

As Don Mike would say. "Sometimes it is simply best to step aside and cheer a community on as they take matters into their own hands. More times than not, you will be amazed."

Don Mike at his computer which frustrated him to no end

CHAPTER SEVENTEEN
Don Mike's Report Card after Ten Years

*"The day you stop learning should be the
same day they put you into the ground."*
Michael Shawcross

It had been a full ten years after the completion of the water project at La Garrucha and I was part of the EWB team that would review the program. Oh, how I had wished that my mentor and friend Don Mike was with me, but he had passed away nearly four years earlier from a massive stroke.

For years, his friends and I kept begging him to treat his high blood pressure. Finally, we had dragged him to the doctor for his first checkup in decades. The doctor was shocked by his high blood pressure and prescribed medication with the strict instructions to begin taking the pills immediately.

"I hate pills and I think they are a bloody waste," he'd said, "They also make me feel shitty."

In 2005, Adam, the author and Don Mike behind three of Don Mike's favorite Highlands teachers

We tried to convince him to change his diet and cut back on his drinking – all to no avail.

"Would you guys please stop nagging me," he would say. "Don't you know you can't teach an old dog new tricks?"

We knew his blood pressure was getting worse as his severe headaches became more frequent. Each of my trips now included several large bottles of aspirin and he was eating it like candy. Finally, the stroke came. He lay in his bed for two more days fighting for his life until a second stroke mercifully took him in his sleep.

News spread quickly throughout the Highlands of Don Mike's death, and everyone was deeply saddened. His sister back in England unexpectedly received a solicitation while she frantically looked for a cemetery that would bury a British Citizen in Guatemala. The solicitation contained countless thumb prints, acting as signatures, from the communities in which he had worked. The communities asked if his sister would allow them the honor of burying Don Mike in the Highlands cemetery in Joyabaj, and she agreed. On the day of the funeral, people from across the Highlands traveled to the cemetery to show their respect and love for the man who had done so much for them.

After Don Mike passed away, the question of what to do with the Landcruiser needed to be addressed. He had failed to draft a will or sign over the title before passing away so the paperwork of transferring ownership would cost more than the value of the vehicle. We knew that we could not simply take it to the junk yard knowing that Don Mike would never forgive us.

One day, a farmer who was working one of the few flat fields in the area, lamented that a tractor would make all the difference in his family's life. After testing out the Landcruiser's ability to pull a plow in the field, it was given to the farmer to be used as an off-road tractor. It continues with its work to this day, grunting and groaning up and down the field rows helping to feed its new owner's family.

Cleaning out Don Mike's home was a chore that nobody wanted to assume. Fortunately, a few of Don Mike's friends grouped together and reluctantly accepted the monumental task of sorting through all of his books and publications. The process took almost a year, and more than a few dumpsters were filled in the process. Many of the books had considerable value and were sold to universities and even the Library of Congress. The funds raised from the sales of the books was spent on several school projects in his beloved Highlands region.

I thought of Don Mike often after his death. I missed his advice and wisdom, but more than anything, I missed his friendship.

The La Garrucha Review.

The review team consisted of two of us who were familiar with the project, Mincho and me, and two EWB Guatemalan staff, Jose and Cata, who would bring a fresh, outside and unbiased perspective. This was a best practice we had learned from Don Mike.

Jose is Quiche Mayan and has worked with Guatemalan communities for over a decade. As the EWB community coordinator for the region, he has worked with scores of communities on assessments, implementations and evaluations.

He is slight in build, but with a quiet strength within. As is typical for Guatemalans, he is reserved and quiet in nature but has a friendliness that people are naturally drawn to. Jose's wife is Kaqchikel Mayan, so he is well versed in the cultural differences of the two Mayan nations – a difference his wife reminds him of everyday. He loves his job, especially working with the Highlands people, and is good at it.

Cata is a powerhouse of a woman packed into a slight frame as she weighs less than a sack of cement. Her job at EWB is Community Engagement, specializing in engaging women. She is a Quiche woman who also speaks fluent Spanish and English. She learned her English while in Tennessee, so her speech has a lovely southern accent that is unexpected when talking to a Quiche Mayan woman.

Cata is always positive and friendly and seems to never have a bad day. Her love of cooking and friendly personality opens doors easily as she enters a kitchen and engages the women of the family in conversation while patting out tortillas. Her skills have proven time and again to be irreplaceable as she unwraps the information given from the women of each family.

Over the years, I had come to refer to her as EWB's "secret weapon." The men would meet with the water committee and she would be helping the women in the kitchen. After the meeting, we would discuss what we learned from the water committee meeting.

Then Cata would say, "Now, let me tell you what is *really* going on." She would then give us the "inside scoop" from her discussions with the women of the community.

Jose was the review team leader based upon his past experience. The EWB review process was based mainly upon the many lessons learned from reviews done with Don Mike. A checklist had been developed and I carefully reviewed it, worried that I might have forgotten one of his best practices.

We had "surprised" the community by visiting them with only a few days' notice as Don Mike had taught me. This was done to prevent any last-minute repairs before our arrival that might skew the results. The data collection and interview questions had been carefully crafted and reviewed by the team before we arrived, and everyone knew their roles and responsibilities. We were ready.

I remember rolling out of my bedroll and rubbing my eyes. The space on the floor had been graciously provided by the community of La Garrucha to the EWB-USA review team and the other team members began to stir. It was the first day of the review and we were anxious to get started.

But I was stopped dead in my tracks by the smell of the fresh Guatemala coffee brewing. I had already heard the mill running at first daylight, grinding the corn into masa for the fresh tortillas. There will be plenty of time for work, I thought. Now is a perfect time for breakfast! As the review team enjoyed our breakfast of coffee and tortillas, we reviewed the plan for the next few weeks with the water committee.

Mincho, two members of the water committee and I would spend the nearly two weeks hiking the water lines and checking the system's mechanics.

Jose and Cata would visit each of the 213 homes with one of the four sector plumbers. The fact that they were also Quiche and spoke fluent Quiche and Spanish would aid them in gathering the information needed. Don Mike always stressed the importance of having a Maya woman on any assessment or review team to assist in the interviews with the community's women.

They would interview the families while verifying the flow and pressure of the water at each tap. Afterwards, they would review the data from the health clinic and school to compare it with that collected ten years ago as a baseline.

Few organizations had such an opportunity to review an infrastructure program after ten years of operation. Unlike other types of community development, infrastructure programs take time to realize their full potential. But they also struggle to sustain themselves over time, especially potable water projects. Most organizations lack the funding and willpower to do such reviews. Funders also tend to shy away from such efforts, choosing to instead fund new projects and hope that they might be sustainable.

I remembered the direction of Cathy Leslie, EWB's Executive Director, to find the real answers that wait to be uncovered. I was determined to not fail her, no matter what the results.

Field Results

Mincho still knew the system like the back of his hand as we walked the lines. He had personally supervised the installation of each and every length of pipe, not to mention all the valves and tanks along the way. I had brought a set of plans and knew the design flows and water pressures at each point which we checked with our meters. Water samples were also collected and checked to verify the water quality of the system.

As was Don Mike's method, we started at the spring and worked our way through the system following the water flow. We verified that the spring flow was the same as ten years ago and then walked the six kilometer conduction line, checking its condition. When we measured the flow at the distribution tanks, it was the same as the flow at the spring, signaling that not a drop of the precious water was leaking.

We all remembered one of the favorite sayings of Don Mike, "Water leaks make me crabby" and laughed knowing he would be smiling.

We had then moved onto the distribution system, noting the adjustments the plumbers had made to the system over the years. Although I was concerned about some locations that now had excessive pressures, the system was fully charged and functioning with only three of the taps (2%) not having water at the time of the visit. At each of the three homes, the plumber was embarrassed and made apologies. He then quickly made arrangements to have the repairs made within the same day.

Clearly, they took pride in the quality of service they were providing.

Meanwhile, the rest of the EWB review team, Jose and Cata, along with a plumber, visited the 213 homes being served by the system.

Ten years earlier, the system serviced 157 homes, so fifty-six new connections had been made to reach the current total of 213. A total of 161 family interviews were conducted. Cata and Jose visited each home's latrine and found that fourteen (9%) lacked a well-maintained latrine at the time of the site visit.

"If I wouldn't want to use that latrine, it is not well maintained," Cata said, explaining her criteria.

As they conducted the interviews, they heard the families talk about how the waterborne illnesses were virtually eliminated with only a few families having a single case of diarrhea in the last three months. This was in stark contrast to ten years ago when the number one complaint by the families was illness due to diarrhea.

They also noted that all the water provided by the system was being used and the distribution tanks never overtopped. Water is a precious resource and it was not surprising to find the uses expanding beyond those that are needed for people. As they visited the homes, they found irrigated gardens growing carrots, squash, tomatoes, broccoli and other vegetables. Many of the homes also had added pipelines to water livestock such as pigs and cattle to take advantage of all of the precious water.

With the community's growth and additional demands on the

water supply, we were concerned that a shortage would likely occur soon. We discussed how water meters could help track the individual usage and provide the option to charge for the water based upon the amount used.

"I now know why Don Mike kept asking us to consider water meters ten years ago," said Rolando.

School Records

The interview team then caught up with Abraham who was now the primary school director, having been promoted from his duties as the first-grade teacher. He explained that the health and hygiene classes for students and adults continue to be offered and well attended. He had also expanded the class topics to include solid waste collection and disposal. He so wished Don Mike had been there to see the students, now gathering the trash in the area, even separating out the organic material for composting to be used in the school's garden.

The school absences due to diarrhea were dramatically reduced. The attendance records were reviewed, and the other teachers were consulted. Currently two student absences were occurring per month due to diarrhea, compared to the more than 300 that had occurred ten years earlier. Abraham also indicated that the students were doing much better in their studies as well, since they were spending more time in class and less time being sick.

"Who wants to be sick? These kids want to learn!" Abraham said to me.

Health Care Records

We learned that the clinic was now testing the water quality twice a year to confirm its safety and proper maintenance. The clinic also provided the chlorine to the plumbers to clean the tanks on a monthly basis. This not only provided the supplies needed, it also was an important check that the work was being done.

The interview team also met with Efrain, Gavina and the other midwives. They explained that most of the mothers were now comfortable going to the regional hospital to give birth, thanks to the encouragement of the Community Health Committee. They showed the records that indicated that no mothers had been lost during childbirth (maternal deaths) in the last ten years compared to the average of one death per year prior to the water system being installed.

Gavina reflected back to the time before the projects. As she thought about the mothers who were her friends and neighbors, that she was not able to help survive childbirth, said softly, "I only wish it could have happened sooner. So many lost their lives needlessly."

When the infant mortality records were reviewed, one infant death had occurred in the last ten years compared to the average of 2.5 per year prior to the water system.

When the health team was congratulated on these remarkable achievements, Efrain said, "One infant is still too many." He shook his head in disappointment.

Clearly, the health team took its role in the community very seriously.

Reflection on the Results

As the review team met afterwards, we discussed what we believed to be the key reasons for the community's improved health. Clearly, the program had achieved its goals of improving the community's health, but Cathy Leslie also wanted to know why. How could EWB learn from the best practices and implement them on other programs?

Identifying and supporting the community's leader is key and Rolando and Gavina's leadership clearly made the difference. Don Mike had identified Rolando early on and supported him throughout the project. Without Gavina's continuous efforts to obtain clean water and the bridge, who knows if the projects would have occurred. As a mother, she clearly has had a strong influence on her son, Rolando.

Not only did Rolando take the risk of leading the initial development of the projects, he also had the vision to identify problems before they became systemic. The way he proactively dealt with the law that impacted the collection of water fees was critical to sustaining the system. All 213 homes had paid their water bill up to six months ago and only eight percent were in arrears for the current monthly bill.

Community ownership of the maintenance and operations is critical, and the volunteer plumbers were remarkably dedicated and professional. They clearly took pride in the system and its service since only three (2%) of the taps were not functioning at the time of the review and no interviewers complained about poor maintenance. Volunteer plumbers are many times a cause of failures with other water systems, but that was not the case here. The old saying of "adversity only makes you stronger" may have applied to the changes required due to the new municipal law that did not allow water to be turned off to the homes. It forced the higher level of community engagement and participation in sustaining the water system.

The team credited the community's participation and full engagement in the water committee as the difference – a trademark of a Don Mike project.

Engaging the community's education program was another key to the program's success. Abraham's health and hygiene education programs were clearly popular and well received by students and adults alike. Abraham was able to make learning fun using pictures and songs which made it all the more effective.

"Teaching is the profession that teaches all other professions," Don Mike would always say and clearly Abraham was an amazing teacher. Don Mike always insisted on partnering with the school on any project to tap into its educational expertise. Once again, it paid dividends at La Garrucha.

Gavina on the right with her friend and
fellow midwife Josepha on the left

But the review team was the most impressed by the Community Health Committee, which the community had started completely on its own. It all started with the midwives and their dedication to improving the health of their neighbors.

The midwives are also a key component of reinforcing the health and hygiene practices in the community's homes as they provide positive reinforcement as well as improvement suggestions with each house call. The fact that only fourteen (9%) of the homes lacked a well-maintained latrine at the time of the review visit must be attributed to the constant vigilance of the midwives.

Efrain's leadership was also key as he saw the importance of building a health care team with the midwives. His advocacy for allowing the midwives to assist in the delivery room at the hospital was critical to removing the obstacle of mothers being scared to go to the hospital for deliveries. The vision he had to form the Community Health Committee by adding in community members was clearly a difference maker. It engaged the community in its own health care solutions and ensured that even the most vulnerable in the community were not left behind. I knew that Don Mike would be thrilled at this example of community development at its best.

We felt another key to the success was the overall structure within the community. The community had formed a system of checks and support by having the Water and Health Committees doing their important roles but also being held accountable by each other and the community. I thought about the different roles and safety nets:

- The plumbers were on the front line doing the regular maintenance and repairs. They were held accountable by the community members on the water committee and the Community Health Committee.
- The school provided education to the children and adults alike with the support and oversight of Efrain and the Community Health Committee.
- Health care was provided by Efrain, Gavina, and the other midwives with the support and oversight of the Community Health Committee and the community.

Everyone was committed to the same vision of a healthy community and was determined to do their part, and support the others, to achieve that vision. They had referred to the previous lack of clean water and a bridge as "the barrier". Once "the barrier" was removed, the community ran with the vision and would not be denied.

From left to right, The Author, his wife Cathy, and Don Mike

Don Mike's Report Card

The review took place four years after Don Mike had passed away and I wished he had lived to see how the community of La Garrucha had taken on its challenges and made such dramatic improvements in its health.

Don Mike had his own Monitoring and Evaluation methods that he had developed over the years. He would visit each of the projects he coached every five years and, true to his passion for education, give them a letter grade.

"It's not just about the bridge, school or water system," he reminded me so long ago. "It is about the project process that empowers communities and its leaders to solve their problems."

I then sat down and attempted to grade the project according to my mentor's grading system.

C Grade: This grade is reserved for projects that are in the same condition that they were built. Clearly Don Mike was a tough grader, as any project that no longer provided service to the community in the same manner as the day it was built would receive a D or F. But La Garrucha met this mark with the bridge that still provided important access to the hospital, schools and markets and the water system that provided clean, safe water to 98% of the users.

Rio Motagua Bridge proudly holding up a
truck load of corn and workers

B Grade: This grade is reserved for projects that met the requirements of the C grade and the community had implemented additional improvements on their own. I smiled, knowing that the Community Health Committee and Abraham's health and hygiene classes clearly met this requirement.

A Grade: This grade is reserved for projects that met the requirements of the B Grade and also the leaders of the project had been recognized, going on to provide leadership inside and outside of the community.

Rolando's leadership was recognized by those in the municipality and he is now an employee of the municipal government, helping to identify and deliver the infrastructure projects in the region.

Abraham's leadership was recognized as he was promoted to Primary School Director due to his dedication and creativity.

Efrain and Gavina's leadership were recognized as they now conduct training classes in western and traditional medicine for all the regional health care workers.

As for the EWB teams, all of the Marquette students graduated and became professional civil engineers, solving infrastructure projects that protect the health, safety and welfare of the public around the globe. Jack is a member of the US Army and has led humanitarian projects around the world. Katie has dedicated her career to tackling the economic challenges of the poor, working in South Africa, Colombia and the inner cities of the United States. James volunteered with AmeriCorps for three years after graduation, creating affordable housing.

All of them have continued to be committed to service work in their own way. Their work continues to be shaped by the experience of working with the people of La Garrucha.

Amy said, "I have the picture of me and Don Poncho's son

Ramone on my desk and look at it every day. It reminds me to look beyond the numbers in my work and see the faces of those who will have better lives because of it."

Amy with Ramone

I could visualize Don Mike looking down from above and smiling. He would be sitting in the Landcruiser and tipping back a glass of his favorite *cusha* to celebrate the community's success. But in the end, he would have given the program a grade of A-, because, in his words, "You should always be striving to do better."

Michael Shawcross "Don Mike" and the author in front of a photo of Don Mike receiving his OBE from The Royal Highness Prince Charles

Epilogue

I continue to visit Don Mike's grave every time I am in the area. I shudder to think what he might think of this book. Surely, as a "Book Man," he would have shaken his head at the grammar and would not have been able to resist himself to edit the text to perfection. But I like to think that he would have reveled in the thought that some of his lessons had been passed forward to future volunteers. He never cared who got the credit, only that others were being helped.

One of the "lovely unintended consequences"—as Don Mike might say—was how the book has reconnected all of those involved in the program so many years ago.

The student design teams, now accomplished engineers, soon rekindled their relationship and the stories flew across social media. Don Mike would be so proud of all of them and how they have progressed with their lives and careers.

Clearly the highlight for me was reconnecting with the people of La Garrucha. I was a bit fearful of how they might react to the book, but they embraced it with open arms.

"We are so proud that our story is being told in a book," Don Rolando said with a smile.

As we shared stories about the past, Florinda, Don Rolando's wife, slipped away back into the house. She reappeared with the family portrait taken twelve years ago during the planning trip for the water project. She had taken the print to the city and had it enlarged. It still hangs proudly over the family's kitchen table.

Rolando and Florinda holding the family photo taken twelve years ago.

I stood in amazement at Rolando's daughter, Rosa. Fifteen years ago, she had handed me a precious egg in the bodega during Hurricane Stan as a young, shy girl. Now she handed me her newborn nephew to hold. Her round, freckled face beamed with a smile as the baby cooed in my arms. The baby was clearly the center of the family's attention and had wrapped his grandfather, Rolando, and his great-grandmother, Gavina, around his short, but amazingly strong little fingers. I could not help but think that with such great lineage, the youngster would be destined for great things.

The community insisted that we have a barbeque on the banks of the Rio Motagua in plain view of the bridge. The chickens of the community quaked once again knowing that they would be called upon to do their part by supplying the tasty grilled meat. I smiled as the pat, pat, pat of the tortilla makers rang out across the valley and the smell of the fresh coffee once again filled my nostrils.

I came to learn that the only story that grows more than a fish story is a bridge-building story. The tales of the bridge construction had grown over the years and continued to be enhanced more and more as we all laughed and stuffed our bellies with tortillas and grilled chicken. With passion, the men insisted they had worked all night during the construction and each recanted stories of their personal feats of strength. This, of course, was helped by a healthy supply of *cusha*. I even started to wonder if another bridge had been built in the community that I was not aware of, as my memory did not recall many of the antics. When I mentioned this to my newly reacquainted friends, they all reassured me that my memory had failed me in my old age and the stories were indeed fact.

As the memories poured from us, our only regret was that Don

Mike was not there to act as the judge and jury to determine who really makes the best *cusha* in the land – once and for all.

As I pried myself away from my friends, Rolando handed me several pages of paper. The community had asked that the following words might be included in the book.

During the years 1981 until 1985 here in La Garrucha, we had very difficult times. We were affected by the violence of the civil war. Our parents were killed and lots of family members were disappeared, and we never saw them again. During that time, we suffered hunger, lack of sleep, illnesses, and injuries. We had to rely on each other to treat one another and we used home remedies to cure our sickness.

For us, the neighbors of La Garrucha community, the construction of the vehicular bridge by EWB USA has been of great benefit to us, in health, access to market and we have more commercial activities. There are many people who now are able to go from Joyabaj to Chimaltenango via San José Poaquil.

The EWB USA members have helped us in the construction of a water system that has been very beneficial for us. Trying to get this project done was a process of 15 years, that was a struggle. We feel very happy that our project was installed 10 years ago. Thanks to God and to the EWB members, our water project is working well.

We please request that EWB continues helping other communities that are in need, just as we were.

If you are so moved to support these efforts, you can make a donation at www.ewb-usa.org or www.milwaukeerotary.com.

May 2020

We find ourselves in midst of the world's worst pandemic in over a century. My thoughts are frequently with the community of La

Garrucha. They assure me that they are as prepared as they can be, but like all of us, they are afraid of the impact COVID-19 will have on them and their neighbors.

As part of a wider effort across Guatemala, Rotary and EWB-USA is helping local manufacturers make Personal Protective Equipment (PPE) for health-care workers. The Marquette students took up a collection and provided the PPE to Efrain, Gavina, Josepha and the other health-care workers. The Health Care Committee is on the lookout for COVID-19 signs and continuously reinforcing social distancing and handwashing. All of the systems established over the last decade – access to the hospital, clean water, Efrain's clinic, the health committee – will be tested by the unrelenting virus.

I am particularly concerned about Gavina and the other midwives. All are advanced in age and fall into the category of those "most vulnerable" to COVID-19. But babies don't stop for the virus, and their mission must continue. They, like so many health-care workers around the world, will not abandon their patients, friends and neighbors in this time of need.

I light a candle and pray for their safety.

Acknowledgments

I would like to thank the Highlands People of Guatemala who have been so willing to share their lives with me. Especially, my sincere thanks go out to the people of La Garrucha who taught and gave me so much – with their most precious gift being their friendship.

Many people have asked me where my volunteer spirit came from. Clearly, it started with my parents, Robert and Joyce. I cannot think of a time when they were not volunteering their time and talents to a community project.

I remember when my father made the decision to volunteer as an Emergency Medical Technician (EMT) in addition to his full-time job as a forester. He would travel three hours round-trip in northern Minnesota to attend his classes on how to help those in need. Between classes, he would practice clearing airways and bandaging wounds on me and my siblings – much to our delight.

His volunteer work as an EMT showed me that volunteering can be hard and requires sacrifice, as many a fishing trip was deferred when he was on duty. Since we lived in a very small town, many of the EMT calls were to help his friends and neighbors, who he faithfully patched up. Sometimes he needed to provide comfort to a friend during their last breath on this earth – a difficult but incredibly important task.

Thank you, Mom and Dad, for putting me on the path of volunteer service.

Thank you to Bernard Amadei and Cathy Leslie, whose vision and leadership have guided Engineers Without Borders USA into an organization that continues to change lives both in the USA and abroad. I would also like to thank the EWB-USA staff, especially those in Guatemala who have grown to be some of my best friends. Your patience with me when "the engineer" would freak out is greatly appreciated.

Thank you also to John DeDakis, who coached me and edited the book. Shannon and the Team at TEN16 Press are amazing, and your flexibility and willingness to undertake this crazy project is most appreciated.

I would also like to thank Scott Mitchell, a longtime friend of Don Mike who helped in supplying many of the photos in the book.

A special thanks goes out to Judy Haselhoef and Avi Lank, who encouraged me when I felt this project was simply not possible. Your faith in me is inspiring. Also, a special thanks to Susan Barnett who rekindled my interest in writing during our trip together in Ethiopia and sparked the idea of this book.

Finally, I must thank my wife, Cathy. She has stuck with this crazy engineer through thick and thin. I can't imagine going through this life without you.

About the Author

Michael Paddock grew up in northern Minnesota and attended Michigan Technological University, receiving bachelors degrees in civil engineering and surveying. He is a licensed professional engineer and surveyor whose professional career at CH2M Hill was spent managing teams of over 100 engineers designing infrastructure projects exceeding $1 billion, and he was the youngest-ever recipient of Wisconsin's "Engineer of the Year" award.

After a near-death cancer experience, he was motivated to begin a pro bono engineering career that has delivered over 100 projects with Engineers Without Borders USA, Rotary International, and other nonprofits on five continents over the last 20 years. He currently lives in southeast Wisconsin with his wife Cathy.

You can stay up to date with Michael's work on his website: bridgingbarriers.com.